SURVEYING THE 120TH MERIDIAN AND THE GREAT DIVIDE

Caitlin Press Inc.
8100 Alderwood Road,
Halfmoon Bay, BC V0N 1Y1
www.caitlin-press.com

Copy edited by Catherine Edwards
Text design and cover by Vici Johnstone
Front cover image: Courtesy Library and Archives Canada PA-018406

Printed in Canada

Caitlin Press Inc. acknowledges financial support from the Government of Canada and the Canada Council for the Arts, and the Province of British Columbia through the British Columbia Arts Council and the Book Publisher's Tax Credit

Library and Archives Canada Cataloguing in Publication

Surveying the 120th meridian and the great divide : the Alberta/BC boundary survey,

   1918–1924 / Jay Sherwood.

       Sherwood, Jay, 1947- author.

Includes bibliographical references and index.

Canadiana 20190132620 | ISBN 9781773860091 (softcover)

LCSH: Alberta—Boundaries—British Columbia—History—20th century. | LCSH: British Columbia—

   Boundaries—Alberta—History—20th century. | LCSH: Alberta—Surveys—History—20th century. | LCSH:

   British Columbia—Surveys—History—20th century. | LCSH: Surveying—Alberta—History—20th century. |

   LCSH: Surveying—British Columbia—History—20th century.

LCC FC208 .S34 2019 | DDC 971.23/02—dc23

SURVEYING THE 120TH MERIDIAN AND THE GREAT DIVIDE

The Alberta-BC Boundary Survey, 1918–1924

Jay Sherwood

CAITLIN PRESS

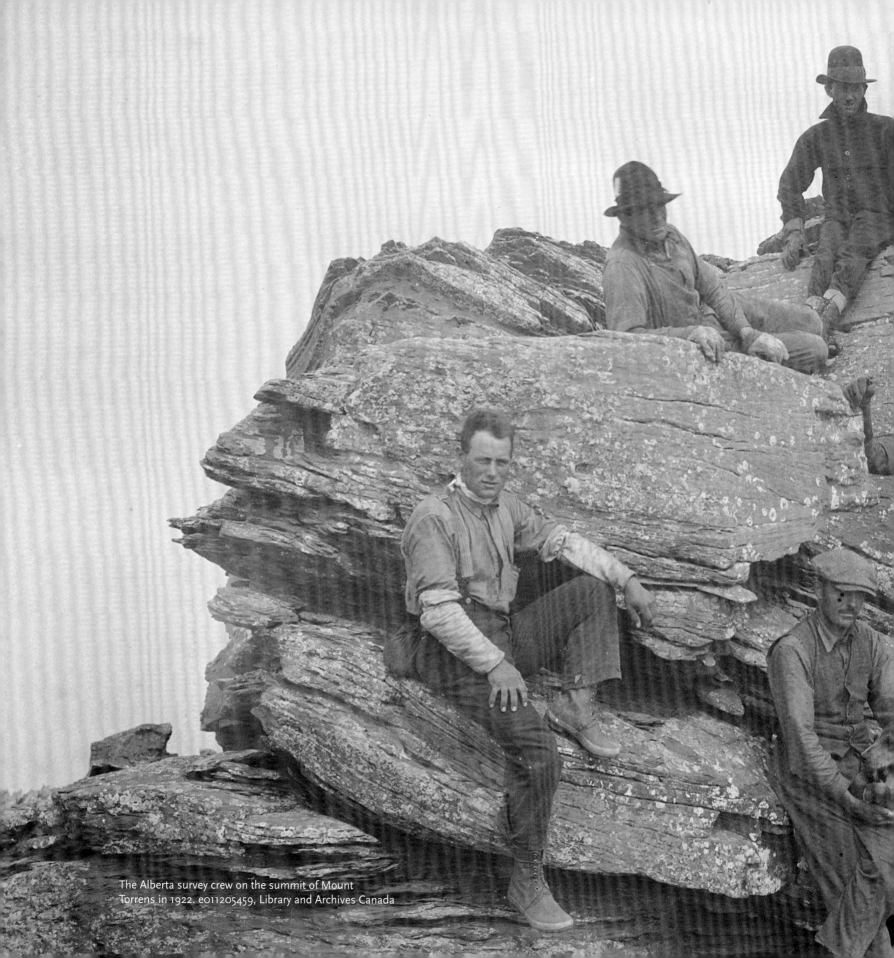

The Alberta survey crew on the summit of Mount
Torrens in 1922. e011205459, Library and Archives Canada

This book is dedicated to all of the people who worked on the Alberta-British Columbia boundary survey from 1918 to 1924.

S-5644

CONTENTS

MAP 1

Map 1

Approximate location along the 120th meridian surveyed by Cautley between 1918 and 1924

Map 2

Approximate location along the Great Divide surveyed by Wheeler and Campbell between 1918 and 1924

Legend

Passes surveyed by Cautley

Year	Name of Pass	Letter
1921	Fortress	O
1921	Athabasca	P
1921	Whirlpool	Q
1921	Tonquin	R
1924	Miette	T
1924	Robson	U

(Yellowhead Pass, S, surveyed in 1917)

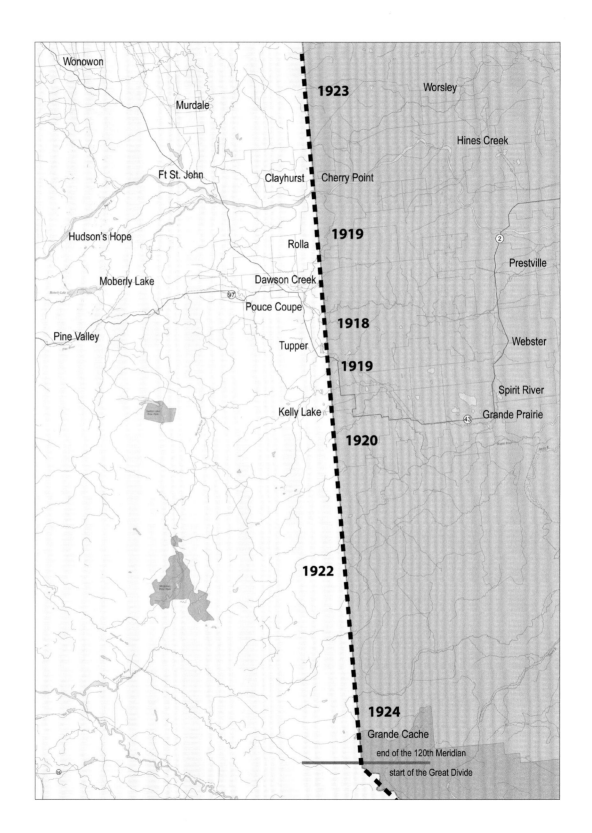

MAP 2

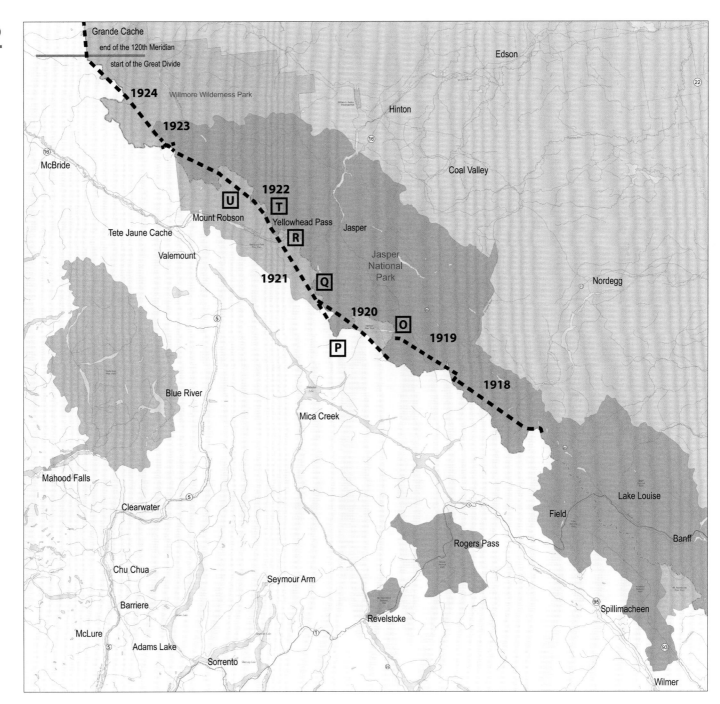

Grande Cache
end of the 120th Meridian
start of the Great Divide

1924
Willmore Wilderness Park

1923

Edson

McBride

Hinton

Coal Valley

U 1922 T
Mount Robson Yellowhead Pass
Tete Jaune Cache R
Valemount Jasper

1921 Q

Jasper
National
Park

Nordegg

1920
O
P 1919

Blue River 1918

Mica Creek

Mahood Falls

Clearwater
Lake Louise

Field
Banff

Rogers Pass

Chu Chua

Seymour Arm

Spillimacheen

Barriere

McLure Revelstoke

Adams Lake

Sorrento

Wilmer

MAP

9

ACKNOWLEDGEMENTS

I would like to thank the many people and organizations that provided assistance for this second and final book on the Alberta-BC boundary survey. Mike Thomson, surveyor general of British Columbia, provided support throughout this project, along with the Association of British Columbia Land Surveyors. The Land Title and Survey Authority made their material available, and I would particularly like to thank Dave Swaile, Calvin Woelke and Tracy Nguyen. Eric Higgs, director of the Mountain Legacy Project, furnished the project's Wheeler pictures and rephotography images from the 1918 to 1924 portion of the boundary survey. Carla Jack, BC's provincial toponymist, provided important information related to the naming of geographical features along the boundary. Buddy Syrette located Wheeler's correspondence files with the BC surveyor general.

My thanks go to the staff at the various archives I visited for their courteous assistance. This includes the Whyte Museum, Jasper-Yellowhead Museum & Archives, Provincial Archives of Alberta, BC Archives, and Library and Archives Canada (LAC). I would particularly like to thank Jean Matheson at LAC for her work with the visual images.

Individuals who helped me include Barb and Mike Helm, who once again gave me a home away from home when I was in Ottawa, a kindness I have always appreciated. Mary Andrews, whose father, Gerry, was BC's boundary commissioner from 1952 to 1968, loaned me some of her father's material, including his annotated copies of the Boundary Commission reports. Thanks again to Robert Allen for sharing his vast knowledge of surveying. Jack Stelfox's descendants provided his recollections of working with A.O. Wheeler in 1921. My thanks also go to Ross Peck for the resources he found. Rob Watt provided assistance in locating R.W. Cautley's diaries.

I would like to express thanks to my family for their support. Once again my wife, Linda, helped me in a myriad of ways. My thanks also go to Vici Johnstone of Caitlin Press for publishing the boundary survey books.

Surveying on a mountain between Banff and Jasper. The surveyor is kneeling beside the mountain transit, which has some rocks in the sack for stability. The man on the right is holding a flag used for signalling to another crew on a nearby mountain. There is an ice axe behind him. PA64-26, Jasper Yellowhead Museum & Archives

INTRODUCTION

The Alberta-British Columbia boundary is the longest interprovincial boundary in Canada, extending for 1842 kilometres from the 49th to the 60th parallels. It consists of two parts: the Great Divide (more commonly called the Continental Divide today), a natural feature along the watershed of the Rocky Mountains; and the 120th meridian, a man-made feature.

The initial survey of this boundary started in 1913 and continued until 1924, delineating the boundary from the 49th to north of the 57th parallel. It was a unique Canadian project that combined talented surveyors, high-tech surveying equipment, rugged crew members and Canadian wilderness. It's also a story of adventure and danger—climbing mountains and surveying from the peaks of the Canadian Rockies; slogging through the muskeg north of the Peace River; occasionally crossing rivers at high water; and often working in the rain, snow or cold.

The survey began at the end of an economic boom in western Canada and involved the Dominion and the two provincial governments. The expenses for the survey were divided equally among the three jurisdictions. Initially, all three governments had their own representative on the Boundary Commission, but in 1915, to save money during World War I, R.W. Cautley, the Alberta member, also became the Dominion representative.

Édouard-Gaston Deville, the surveyor general of Canada, provided overall leadership for the project. Deville was one of the world's leading authorities on phototopographic surveying, and the Alberta-BC boundary survey was the longest and most extensive project (in years) of this type of survey in North America. Each province had responsibility for one survey crew.

R.W. Cautley was in charge of the Alberta crew. From 1913 to 1917, he surveyed several passes through the Rocky Mountains. Cautley defined the boundary in these passes with concrete monuments, many of which still survive a century later, and he also produced detailed maps of these areas. At that time, Cautley's work was considered the most important part of the survey because these passes were economically and geographically significant. Both provinces also needed to know the location of the boundary through the passes for jurisdictional purposes.

A.O. Wheeler, a surveyor and president of the Alpine Club of Canada, headed the British Columbia crew, which surveyed and mapped the topography of the area along the peaks of the Great Divide, using a technique known as phototopographic surveying. It consisted of two parts: triangulation surveying and a series of photographs taken at the survey stations. The two components were used in combination to produce the first detailed, accurate maps of the Great Divide.

From 1913 to 1916, the Alberta-BC boundary survey covered the Great Divide from Kicking Horse Pass (west of Lake Louise) south to the international boundary with the United States. In 1917, the two crews surveyed Howse Pass, the first major pass north of Kicking Horse, and Yellowhead Pass, the largest pass on the northern part of the Great Divide. Wheeler's crew also finished the phototopographic survey between the Kicking Horse and Howse passes. The Boundary Commission report stated:

> From the inception of the Interprovincial Boundary survey in 1913 until the end of 1917, the Commissioners [Wheeler and Cautley] have worked in more or less close conjunction. There are a number of important passes to the south of Kicking

R.W. Cautley's crew preparing to depart from Grande Prairie at the start of a field season. PA-018401, Library and Archives Canada

Horse Pass, the survey of which made it necessary for the Commissioners to meet frequently.

This portion of the Alberta-BC boundary survey is described in *Surveying the Great Divide*, the first book in this two-part series about this project. The Boundary Commission report also commented: "At the end of 1917, however, Mr. Cautley's division had surveyed Howse and Yellowhead Passes while Mr. Wheeler's division was just entering on a long stretch of the wildest and most inaccessible part of the main range, extending northward from Howse Pass." During the second part of the boundary survey, which is described in this book, the two survey crews worked separately.

By the spring of 1918, Canada had been involved in World War I for almost four years. Although it would be the final year of the war, at that time there appeared to be no imminent end to the conflict. Each year, the war had been taking an increasing toll on the country in lives lost, financial costs and stress on society. The expenditures of both the federal and provincial governments were primarily directed towards activities supporting the war and maintaining basic services. But despite the difficult economic conditions, the federal government, along with the Alberta and British Columbia governments, decided to continue the boundary survey between the two provinces. During World War I, the federal government and the two provinces had deemed the project of sufficient importance to continue funding it, although on a significantly reduced scale. After protracted discussion, funding was once again provided for work in 1918, but only at the minimum amount needed to sustain the survey.

After completing the southern portion of the Great Divide, the BC government wanted to start surveying the 120th meridian so that it could promote development of agricultural land in the Peace River region. Before this could begin, the federal government sent F.A. McDiarmid from the Dominion observatory to the Peace River region during the summer of 1917. McDiarmid built a longitude pier and established the latitude and longitude of a point on this station. The Boundary Commission report stated: "Since the only telegraph line which crossed the 120th meridian, north of the mountains, was the single wire Government telegraph line along Fort St. John road, and the telegraph office nearest to the 120th meridian was at Pouce Coupe, the pier had to be established at that point." Using specialized, high-tech equipment, astronomical observations and telegraph signals, McDiarmid spent almost four weeks ascertaining the precise location of this pier.

In 1918, R.W. Cautley began surveying the 120th meridian, and this was the main focus of the Alberta survey crew for the next seven years. Wheeler, meanwhile, continued phototopographic surveying along the Great Divide until he completed the work in September 1924. From 1922 to 1924, Deville sent H.F. Lambart from the Dominion Geodetic Survey to assist Wheeler. There were only a few survey stations north of Yellowhead Pass, so Lambart established a triangulation network that included survey stations that

Wheeler could use to connect with his work. Although all three crews were striving towards a common purpose, their surveys in the field were separate. In this book, therefore, the work of each crew is covered by itself. The final section of the Alberta-BC boundary was not surveyed until after World War II. This book includes a short chapter on the completion of this quintessentially Canadian project.

THE SURVEYORS

The initial Alberta-British Columbia boundary survey lasted from 1913 to 1924. The major characters involved with the project were some of Canada's most talented surveyors of that time period, and all but one of them worked for the entire twelve years of the boundary survey. (J.N. Wallace's position as Dominion commissioner was combined with Alberta surveyor R.W. Cautley's during World War I.) For these people, it was one of the major activities of their surveying career.

ÉDOUARD-GASTON DEVILLE, SURVEYOR GENERAL OF CANADA

Born in France in 1849, Deville began his career with the French navy and was in charge of hydrographic surveying projects throughout the world. He immigrated to Canada in 1874 and was appointed inspector of surveys by the Quebec government the following year. In 1878, he obtained his Dominion land surveyor's licence and started working for the federal government as a topographic surveyor. Three years later, he became an inspector for the Dominion Land Surveys, and in 1882, he was appointed head inspector. In 1885, Deville was appointed the federal government's surveyor general, a position he held for thirty-nine years.

Deville was particularly interested in phototopographical surveying and started applying it to conditions in western Canada soon after becoming surveyor general. He perfected techniques developed by Aimé Laussedat, a French army engineer, and developed equipment that could be used in mountainous terrain. Deville became one of the world's leading experts on phototopographic surveying, and by the time of the Alberta-BC boundary survey, the importance of this technology had been proven. He wrote a book titled *Photographic Surveying* and is considered the father of photogrammetry in Canada.

Declining health limited Deville's direct involvement in the boundary survey during the final years of the project, but he retained an overall supervisory role.

This photograph of Édouard-Gaston Deville was taken around the beginning of the boundary survey in 1913. PA-043041, Library and Archives Canada

RICHARD WILLIAM CAUTLEY, ALBERTA COMMISSIONER, DOMINION COMMISSIONER

Born in England in 1873, Richard William (R.W. or Bill) Cautley immigrated to Vancouver when he was seventeen. In 1892, he began articling to be a surveyor with James F. Garden, a BC land surveyor who later became mayor of Vancouver. Cautley obtained both his provincial and Dominion land surveying licences in 1896. The following year, he was appointed to be one of the surveyors on the gold commissioner's staff for the Yukon Territory and spent six years there. In 1903, Cautley moved to Edmonton, and until 1909, he worked for the federal government making several surveys in the prairie provinces. He also established a private surveying practice with his brother, Reginald. In 1909, Cautley was appointed surveyor for the Alberta government's land title office, a position he held until his appointment as Alberta's commissioner for the boundary survey. Cautley also wrote a textbook for surveying students that was used throughout Canada for many years.

R.W. Cautley in front of his office tent.
A8090, Provincial Archives of Alberta

ARTHUR OLIVER WHEELER, BRITISH COLUMBIA COMMISSIONER

A.O. Wheeler was born in Ireland in 1860, and immigrated with his family to Canada in 1876. He articled to be a surveyor in Ontario and received his licence in 1881. The following year, he became both a Dominion and a Manitoba land surveyor. At the beginning of 1885, Wheeler began working in Ottawa as a surveyor for the Department of the Interior, and he learned about phototopographic surveying from Deville. That spring, during the Riel Rebellion, Wheeler served with the Dominion Land Surveyors' Intelligence Corps, and was slightly wounded at the Battle of Batoche. Wheeler married Clara Macoun, the daughter of a famous Canadian naturalist, John Macoun, in 1888.

During the 1890s, most of Wheeler's work was surveying for irrigation in the western part of the Alberta district of the Northwest Territories. Wheeler's work included the use of phototopography to assist with the maps that he produced. In 1900, he began working on phototopographic surveying projects in both the Selkirk and Rocky Mountain ranges. Wheeler had a keen interest in the outdoors and mountaineering. In 1906, he was one of the founders of the Alpine Club of Canada and elected as its president. During the following summers, Wheeler took time from his surveying to organize and attend the Alpine Club's annual summer camp. Deville told Wheeler in 1910 that he could no longer attend the camps because it was interfering with his phototopographic surveying. Wheeler received an exemption for one more summer, and then he decided to resign and go into private practice.

For the boundary survey, both Deville and the BC surveyor general, G.H. Dawson, selected Wheeler as their first choice to be the Pacific province's commissioner in charge of the topographic and phototopographic surveying. Wheeler agreed, providing that he was allowed to attend the Alpine Club's summer camp. He hired A.J. Campbell, one of the men from his surveying company, to be his assistant and to take charge of the surveying when Wheeler was absent. During the 1920s, Wheeler spent less time in the field, and Campbell assumed responsibility for most of the actual surveying. However, Wheeler continued to do the office work, making the calculations and producing the maps, and he maintained overall supervision of the BC portion of the project.

Alan John Campbell, Assistant Surveyor, BC Crew

A.J. Campbell was born in Ontario in 1882 and grew up in that province. He obtained a diploma in civil engineering and a bachelor's degree in applied science from the University of Toronto. While articling with W.J. Deans to be a surveyor, Campbell first came to western Canada in 1908, working on surveys in Manitoba and Saskatchewan. The next year, he arrived in British Columbia and began surveying for A.O.

A.O. Wheeler with a phototopographic camera. V465-pd3-374, Whyte Museum of the Canadian Rockies

Wheeler. In 1911, Wheeler formed a survey company with Campbell and R.D. McCaw. Campbell and Mc-Caw spent much of that summer surveying a large number of lots in the Tetachuck Lake area of the upper Nechako headwaters.

In 1913, when A.O. Wheeler was appointed BC's commissioner for the Alberta-BC boundary survey, the company dissolved. McCaw began working for the BC government, initially on the Banff-Windermere Highway, and then in 1914, he started the provincial government's phototopographic surveying program, a lifelong project for him. Campbell became Wheeler's assistant, and was in charge of the topographic survey during Wheeler's frequent absences. Campbell also worked for Wheeler during the winter, helping prepare the notes and maps made from the information gathered during the summer's work. Wheeler was known to be a difficult person to work with, but the quiet, soft-spoken Campbell remained for the entire project, and for those twelve years was the person who worked most closely with Wheeler.

During the later years of the Alberta-BC boundary survey, Campbell assumed responsibility for most of the surveying fieldwork.

A.J. Campbell. 89.03.218, Jasper Yellowhead Museum & Archives

SURVEYING METHODS, 1918–1924

The Alberta and British Columbia survey crews used different methods of surveying for different purposes during this part of the boundary survey.

From 1913 to 1916, R.W. Cautley surveyed the main passes along the Great Divide between the United States border and Kicking Horse Pass, west of Lake Louise. In 1917, he surveyed Howse Pass, the first pass north of Kicking Horse, along with Yellowhead Pass, west of Jasper. Through a series of monuments constructed in each pass, Cautley physically defined the location of the divide. (Cautley made a preliminary survey of a pass and then conferred with Wheeler. After both surveyors concurred with the location of the monuments, Cautley would make the final survey and establish the monuments, many of which still exist today, a century later.)

From 1918 to 1924, Cautley surveyed passes only in 1921 and a portion of the 1924 field season. The passes along the northern part of the divide were smaller and more remote. Only the Yellowhead, with the Grand Trunk Pacific and Canadian Northern rail lines, had any economic significance.

The main objective for Cautley during these seven years was to define the Alberta-BC boundary along its northern portion, the 120th meridian north from its intersection with the Continental Divide. As with his work in the passes, Cautley and the Alberta crew cut a survey line and physically defined the 120th meridian with a series of monuments. They initially used the same concrete design. However, Cautley found it difficult to find suitable gravel in many locations. Unlike the Rocky Mountains, the terrain along the 120th meridian had almost no bedrock on which to construct the monuments, so the monument-builders had to establish a full monument at each station, making their work slower and more difficult.

In a letter to Deville in 1919, Cautley wrote:

> There is one point in connection with the subject that makes monument building on the 120th meridian a more arduous business than it is in the mountains, namely that whereas in the mountains a great many monuments—particularly in the higher sites—only require one third of the material used in a full-sized monument, because they are built on bedrock very close to, or even above, the general surface, every monument built on the 120th meridian is of full standard size and requires 3000 lbs [1360 kilograms] of gravel, cement and water for its construction.

Beginning in 1922, changes were made to the monuments Cautley used along the meridian line.

Cautley's work was labour-intensive. Except for 1918, during World War I, he had a crew of about fifteen people each year. Cautley had an assistant surveyor (except in 1918), chainmen for measuring the survey line, a leveller and assistant to measure elevation, and a monument-builder and assistant. In many places along the 120th meridian, the timber was very thick, so several axemen were hired to cut the line. To support the crew, there were cooks and packers.

F.A. McDiarmid from the Dominion Observatory came to Pouce Coupe during the summer of 1917 and established a longitude pier near the community. Using signals from the telegraph line, he determined the longitude at that location. When Cautley and his crew arrived in 1918, they had to make a connecting survey to the 120th meridian. The longitude pier was 8.3 kilometres west of the 120th meridian, so it was imperative to measure the exact distance to this location. The Boundary Commission report stated that

> the connecting survey called for extreme care since any error in its measurement would have resulted in the entire boundary being out of place by the full amount of such error. The connection was made along the north boundary of Township 77 and chained twice by the transit and tripod method, after the chain had been very carefully compared with standard tapes.
>
> As soon as the above survey had been made, concrete Monument 78-1 was built 3.079 chains [61.6 metres] north of the north boundary of Township 77 and forms the initial point from which the survey of the 120th meridian was subsequently carried north and south.

Cautley described his work: "The running of the 120th meridian is a simple survey, depending for its correctness on the accuracy of the astronomical observations, instrument work and chaining of the surveyor in charge and based, in the first place, upon data as to the location of the meridian." He needed to survey south until he intersected the Great Divide and connected with the British Columbia survey. Cautley also had to delineate the 120th meridian line north through the agricultural land of the Peace River region, for there had not yet been a definitive survey, and many people did not know whether their land was in Alberta or British Columbia.

From 1913 to 1916, Wheeler made a phototopographic survey along the Great Divide from Kicking Horse Pass south to the United States border. In 1917, he started the second section, from Kicking Horse to Yellowhead Pass. Wheeler and his crew surveyed to Howse Pass, the first main pass to the north, and then joined Cautley in defining the boundary through Yellowhead Pass and some adjacent terrain.

The surveying methods used by Wheeler and the BC crew remained unchanged during the second part of the survey. The Boundary Commission report stated: "The topographical work of the Commission

Marcus Platz, monument-builder (left), and Harold Clutton at Monument 73-7. e011205469, Library and Archives Canada

is entirely in the hands of Mr. Wheeler. It consists in the delineation of the watershed along the main range from the International Boundary northward to the final crossing of the 120th meridian of west longitude." There were two main aspects to Wheeler's work: the triangulation survey and the photography. In triangulation surveying, if two angles and any side

The 120th meridian line cut by Cautley's crew. e011205436, Library and Archives Canada

their direction established by the use of the transit-theodolite, frequently entailing a precarious balancing over dizzy depths. Rock cairns are erected at the selected stations, either in advance or at the time they are occupied, and are used for identification purposes.

The stations are fixed in position by a triangulation of a greater or less degree of precision, expanded from a given base and extended over the required area. This triangulation may be made independently or be carried on at the same time as the photographing.

Usually the triangulation surveying and phototopography were done simultaneously, for the surveyors did not want to climb a mountain more than once if it wasn't necessary. In addition to the main surveying station, Wheeler or his assistant, Campbell, would occasionally establish a second station that was used primarily for phototopography. Usually this station was at a lower elevation but with a better view of the terrain.

The phototopographic camera was designed to fit on the same tripod as the transit, so that both operations could be completed while working at a survey station. In addition to measuring angles, it was important to determine the elevation of the stations and the geographical features. The Boundary Commission report explained the process.

distance of a triangle are known, the remaining angle and distances can be computed. From mountain peaks, Wheeler would survey to other stations and peaks in the area, creating a network of triangles. While the surveying provided the framework for mapping the Rocky Mountains, the phototopography enabled him to map the details. The report commented that "it was decided to adopt the method of photo-topography as the one best suited to the mapping of the highly accentuated contours of mountain areas."

The report described the process.

> The work requires a specially constructed camera and mountain transit-theodolite. It is carried on by climbing to previously selected stations at the summits of peaks, or to high points on mountain ridges that command a view of the area to be mapped. From these a series of views is taken and

> The altitudes of the various points are obtained by reading angles of elevation or depression back and forth from point to point and applying the necessary corrections for curvature and refraction, a process known technically as trigonometric levelling.

> The chief advantage of the method is its rapidity for work in the field. It is much more rapid than any other that can be applied to the same class of country and the standard of accuracy is a high one. It is true that work can only be carried on in

fine weather, when the stations are below the clouds and the landscape is sufficiently clear to be photographed, but the same limitation applies to any other method that can be used. Only a small party, generally a surveyor and one or two assistants, is required to carry the instruments, take the views, record the transit readings and build the rock cairns or other signals that may be erected to mark the stations it is desired to perpetuate. The instrument outfit, adapted by Dr. Deville, weighs about 45 lbs [20 kilograms], and is so disposed as to be easily carried, even when the climbing is dangerous.

The camera is of fixed focus and has a wide-angle lens covering about 52 degrees of arc for one view. It is adapted in two positions to a light, strong tripod with sliding legs, to which are attached levelling screws to bring the plate exposed in the camera to a vertical position, an absolute necessity to obtain suitable views. The same tripod fits the transit-theodolite. [Each photograph covered about 45 degrees of the landscape with a couple of degrees of overlap on each side. A set of eight photographs would cover the scene.]

The size of the plate used is 6½" by 4¾" [16 by 11 centimetres], and the lens gives very nearly a true perspective. Later, in the office, the plates are developed and bromide enlargements, 9½ by 13 inches [24 by 32 centimetres] are made for mapping purposes. Practically speaking they are perspectives of the views and, by means of geometric and perspective constructions, the position and altitude of points seen in them can be obtained and the contour outlines of the various topographical features drawn on the plan.... In the mapping room, this chaotic condition soon resolves itself into orderly array.

There are certain factors necessary to the conduct of a survey such as this, for

A rock cairn constructed at a station on the top of a mountain. e011205452, Library and Archives Canada

instance: the triangulation on which the photographic work depends must start from a measured base and be expanded in sufficiently symmetrical proportions; the altitude above sea-level of one or both ends of the base must be known; and there must be independent checks on the expansion of the triangulation at sufficiently close intervals to keep the work within bounds of the limit of accuracy.

The maps that Wheeler produced from the photo-topographic surveying were used to delineate the entire

A.O. Wheeler's crew surveying on a mountain. The man on the left is by the transit, while the man on the right is recording data in the field notes. PA64-27, Jasper Yellowhead Museum & Archives

Below: After the angles were read with the transit, the phototopographic camera was used to take a series of pictures at the same station. 89.03.206, Jasper Yellowhead Museum & Archives

Continental Divide from the United States border to the 120th meridian. Unlike Cautley's work, the boundary was not physically marked, although some stations were located on it. In the section between Kicking Horse Pass and Yellowhead Pass, Wheeler and Campbell had very few surveys to connect with their work. In the final section north of the Yellowhead Pass, Deville arranged for the Geodetic Survey to provide assistance for the BC surveyors.

In his 1913 report to the BC government, Wheeler wrote that surveying on the mountains had both a routine and dangerous aspect.

> Outside of the difficulties of carrying the survey instruments to the summits of commanding peaks, a process entailing all the delights and dangers of mountaineering of a high order, after a few years' experience, the work becomes mechanical, except for the glories opened up by each new climb amidst the wonderful alpine scenery displayed by the Canadian Rockies.

The vagaries of the weather in the mountains often produced difficulties. On many occasions, the survey crews would leave camp in clear weather in the morning, only to find that conditions had changed by the time they reached the locations on the mountains where they planned to survey. Wheeler wrote:

> To be perched on a high peak with standing room only when a tempest of snow scours over it is by no means a joy; neither is an electric storm, when your transit sings like a telegraph wire and to touch it gives you a sharp shock, when the rocks around you hum and your hair stands straight on end.

The surveying talents of both Cautley and Wheeler were essential for the success of the Alberta-British Columbia boundary survey.

One man is reading the horizontal angle, one the vertical. Speed and efficiency were important when surveying on the mountains.
A10727, Provincial Archives of Alberta

Steam tractor and wagon on way to Pouce Coupe passing Cautley's camp. e011205439, Library and Archives Canada

1918

In 1918, A.O. Wheeler and the British Columbia crew continued the phototopographic survey of the Great Divide, heading north from Howse Pass, where they had worked the previous year. The rugged terrain and the crew's overall lack of experience probably contributed to some of the mishaps that occurred during the field season. At the same time, R.W. Cautley and the Alberta crew started a new type of survey in a new location along the Alberta-British Columbia boundary.

Both Wheeler and Cautley had to contend with not only a limited budget but also a shortage of personnel in 1918. Almost every able-bodied person was either serving in the war or employed in an occupation that supported the war effort. There were many jobs available for people seeking employment, and the prospect of working for five months in a remote wilderness area did not appeal to most people. Since this was government work, Wheeler and Cautley had to give first priority to returning soldiers, which prolonged the hiring process.

A.O. WHEELER

Spike Thomson, second assistant, and Ralph Rink, head packer—two important members of Wheeler's crew in 1916 and 1917—were serving in the war in 1918. The only returning personnel were A.J. Campbell, assistant since the beginning of the boundary survey, and Tom Martin, cook for the previous two years. Wheeler hired one returning soldier; Lee Grant Cameron was a trumpeter who had enlisted when he was underage. After his actual age was discovered, he was discharged. Wheeler's head packer was Dell Thomas. His assistant was William C. Hobock, a character from the American frontier who had his own business buying and selling horses and mules in the Pacific Northwest before coming to Canada. Fortunately, Wheeler was able to hire A.E. Thompson, a BC government employee who worked in the land title office and had surveying experience.

The crew assembled at Banff. On June 25, they departed on horseback for Field, the first station along the Canadian Pacific Railway in British Columbia; they arrived there two days later. Wheeler, accompanied by his wife, Clara (in his diaries he calls her Clare), and Thompson, took the train to Field on July 1, where Campbell and the survey party met them. The men camped at the same location that Wheeler had used in 1917 when the BC crew started the field season there. In the evening, Wheeler took Clara back to the hotel in Field. "Said goodbye to her and walked back to camp," he wrote in his diary. Clara's health was fragile, and Wheeler was probably concerned for her, so the next morning "after breakfast packed up my tent & stuff and walked to Field. Clare at breakfast. Saw her off on the train which was on time. Party packing horses when I got back to camp."

Wheeler followed the same route to Howse Pass that he had used in 1917. On the second day of travel north, the crew crossed Amiskwi Pass during a light snowfall. In his diary, Wheeler wrote:

> Reached old campground on the Blaeberry [River] soon after 5 pm & pack train came along shortly after. Good campground. Put up my tent in same old spot—comfortable bed. Below summit of Amiskwi Pass on north side two great snow slides have blocked valley, had to work around them.

Wheeler arrived at Howse Pass on July 4.

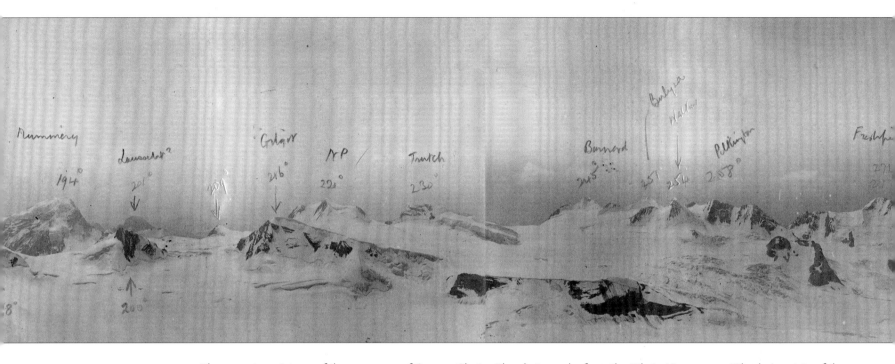

These are two pictures of the panorama of Conway Glacier. The photographs from the Whyte Museum are Wheeler's prints of the glass plate negatives. He put them into his personal photo albums and placed notes on some of the pictures. On these two photographs Wheeler labelled the mountains that he surveyed and the degree so that he could use this information as reference for his field notes. Trutch is labelled on both pictures, so this is part of the overlap of the two photos. The small peak labelled on the left side of the panorama is 188° while Freshfield on the right is 274°, for a total of 86 degrees. The two photographs covered about 90 degrees, a quarter of the full 360-degree panorama. V771_pd_14_4 and V771_pd_14_5, Whyte Museum of the Canadian Rockies

When half a mile [800 metres] from the Howse Pass meadow stopped by a tremendous jam caused by snow slide. Had to take horses up steep hill side to west and went through the brush until we came out at the end of the meadow here to our old campground and got comfortably settled in…. Put my tent in the old place.

The next day was spent getting camp organized and preparing for a fly camp. (A fly camp was a temporary shelter that was used when surveying from a location that was difficult to access or too far away to reach in one day. The cook and packers would usually remain at the main camp while the survey crew carried their equipment and a minimal amount of supplies to the place where they surveyed for a few days. Usually they took only a rain fly for shelter.)

Generally, the Great Divide follows a northwest-southeast direction, but there are places where the summit of the Rockies loops around a geographical feature. After crossing Howse Pass, the divide makes a southwest and southeast loop around the Freshfield Icefield before turning northwest again. To accurately map the boundary through this location, it was necessary to survey two stations. On July 6, Wheeler, Campbell, Thompson and Cameron left for a camp at the tongue of Conway Glacier. The men hiked to the base of a cliff at the foot of the icefall. "After our lunch we packed up the cliff to timber line and made camp in a little point of bush. Snug camp but water a long way down in ravine. Got up cliff fairly well but new hands found it difficult with packs."

Leaving Cameron in camp the next day, the other three men continued climbing the mountain. Wheeler described the ascent.

Campbell, Thompson and I crossed towering icefall and traversed snowfield for about three miles [4.8 kilometres] to a peak of the Divide at S.W. corner of snowfield. Had about 1000 foot [305 metres] rock climb. Reached summit at 12:30 pm. Got a lot of fine views for our work but quite smoky in distance.

During the descent, Wheeler had an accident: "On way down mountain I had a slip on an ice slope and slid down for a considerable distance, fortunately stopping before I got hurt. Went into a crevasse but Campbell got me out with a rope." Since the three men were roped together, Campbell and Thompson were able to keep Wheeler from falling all the way into the crevasse.

On July 8, Wheeler surveyed a second station from their fly camp. He wrote:

Followed yesterday's route for some distance and then struck up northern arm of Conway Glacier and occupied peak at N.W. corner. Good views but still smoky. Returned across snowfield without mishap. Campbell & Cameron with me. Left Thompson in camp. Reached camp about 7:30 pm. Thompson had good supper ready for us…. Felt pretty good for three such hard days. Mosquitos very troublesome. Turned in early. Felt pretty good at having got through so successfully.

In the morning, the men left after breakfast and returned to the main camp by noon.

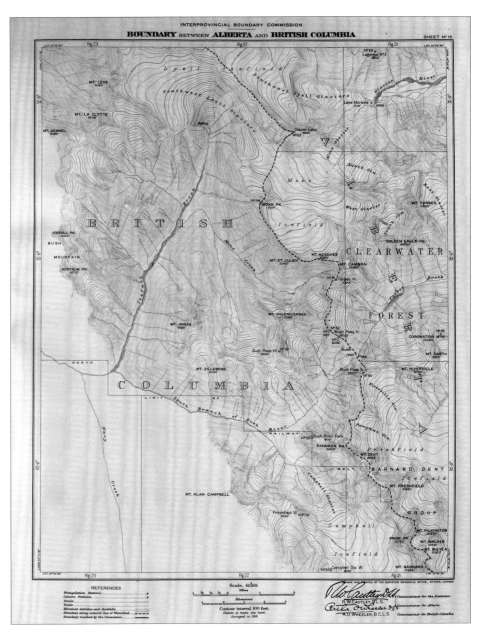

This map covers much of the terrain surveyed by Wheeler during the first part of the 1918 field season. Wheeler named several of the geographical features along this section of the boundary for World War I battles (see Geographical Names section). "Boundary between Alberta and British Columbia" map atlas, Sheet 19, Land Title and Survey Authority of British Columbia

The next objective was Bush Pass, north along the divide. To reach the pass, the crew had to travel north into Alberta from Howse Pass down Conway Creek. In 1917, Wheeler had followed this route while travelling to Yellowhead Pass. When the crew reached Howse River, they followed it upstream for a short distance before proceeding up Forbes Brook. It was raining on July 10, so the men remained in camp. Wheeler and Campbell did some reconnaissance and found an old trail leading to the pass, but it needed to be cleared before they could take horses on the route. The next day, the men cut the trail about 5 kilometres up Forbes Brook, where they located a good, previously used campground. On July 12, the crew moved to this site, which became their main camp in the area. As they were setting up camp, the cook got kicked in the face by a horse—"a nasty cut but not very deep," Wheeler wrote. From this location, Wheeler and three crew members made a reconnaissance trip up the valley almost to Bush Pass. Wheeler described it as a rock saddle and thought that it was doubtful that they could take horses over it. On the way back to camp, the men cleared the trail for most of the distance.

After a Sunday in camp, the crew headed up the valley towards Bush Pass on July 15 and established a camp about 2.5 kilometres below it. Wheeler "went ahead to explore pass & reached foot of wall about 12 o'clock. Grizzly passed on snow about 200 feet [60 metres] away. When he got wind of me he ran like blazes. Explored pass and climbed mountain on north side—got a good look over, made a plan."

On July 16, "Campbell & Cameron took two loads of grub on down the valley to Bush River and found camp ground near toe of glacier on W. side Freshfield group—With Thompson occupied two stations on mountain to north of pass. Did not get finished until after 6. Got to camp at 9 pm, Campbell at 9:40."

The next day, carrying heavy packs, the four men left for an extended fly camp to map the British Columbia side of the divide's loop around the Freshfield Icefield. After a day of rest, the crew surveyed their first station.

> With Campbell and Cameron went up to Campbell Glacier West Sta[tion] via glacier. On way up had a narrow escape from falling ice. Arrived at summit of peak at 11:30 am. Got round of views.… While at azimuth bad thunderstorm came up which lasted an hour very nearly right overhead. When it cleared got good views at second station and also azimuths. Started building cairn but while at it several thunderstorms came along—pretty bad but not quite so bad as first. Managed to get a little shelter and did not get wet. Cleared and finished cairn. Nearly 6 when left hill—got to camp 8:45 pm.

Wheeler attempted to survey another station the next day. In his diary, he wrote:

> Up at 5 and away at 7:15. When at last timber, above moraine, got under trees and made a fire waiting to see what day was going to do—by 10:30 blue sky appeared and day looked somewhat promising. We started up glacier and across snowfield. Travelled up same until 12:00 o'clock, then ate our lunch on the snow. It then clouded over and all hope of getting to a station was gone, so turned homeward. Soon began to rain and we got wet before reaching camp.

On July 21, Wheeler took Cameron to assist him in doing a low station east of Bush Pass. The two men then returned to the camp below Bush Pass. There he found Hobock and the cook and learned that his head packer had cut his foot severely with an axe and returned to the main camp. The next day, Cameron went to rejoin Campbell, while Wheeler left for the main camp in preparation for attending the Alpine Club of Canada's annual summer camp, which was being held in Paradise Valley, near Lake Louise. (Wheeler organized and attended the Alpine Club camps every year. During his absence, Campbell was in charge of the crew.) On July 23, Wheeler and Hobock departed. Following the same route that they had taken into the area, the two men reached Field two days later. At that time Hobock left the crew.

Campbell continued surveying with Thompson and Cameron, and during the rest of July, they occupied five stations on the BC side of Bush Pass. Bad weather hampered their work. The men then crossed back to the Alberta side of the pass, and surveyed a station on nearby Coronation Mountain. In the Boundary Commission report, Wheeler described what happened.

> The party… on the 1st August made the ascent of Coronation Mountain, when there occurred a very serious mishap resulting in the absolute loss of a book of field notes of the work from the commencement of the season. The climbing party in charge of the chief assistant, Mr. A.J. Campbell, made

These two photographs are part of the panorama of the Freshfield Icefield. The small peak labelled 269° is on both photographs. V771_pd_14_16 and V771_pd_14_17, Whyte Museum of the Canadian Rockies

the ascent of the peak by a route both difficult and dangerous and completed work at the summit. When descending by another route it became necessary at one place to lower the other two members of the party—both of whom were novices and in their first season at mountaineering—down a bad piece of cliffs; the survey instruments, camera, transit, and a rucksack containing a satchel with the field book in it, sweaters, etc., then had to be lowered separately. Untying the rucksack from the rope, one of the assistants placed it upon a too narrow ledge and the moment he removed his hand it fell off, struck the ledge on which he was standing and bounded over the edge out of sight. Climbing down it was nowhere to be seen. For two days, in pouring rain, every possible spot in the vicinity was searched and only one conclusion was probable. On bounding over the ledge the rucksack must've fallen into

a narrow rock gully with a steep incline to its mouth, across which flowed a mass of glacier ice. Directly opposite the mouth of the gully there was a large hole in the ice, doubtless carved by water flowing down the gully, which furnished a run-off channel for the melting snows from above. The incline of the gully continued steeply under the ice and the only conclusion possible was that the rucksack had continued its course down the gully and under the ice. Mr. Campbell lowered a weighted rope for one hundred and fifty feet [45 metres] down through the hole in the ice and found that the steep incline continued beyond that distance. There was no possible way of ascertaining where the rucksack had gone.

This serious loss necessitated the re-occupation of a number of stations on the west side of Bush Pass, and Mr. Campbell, who in my absence was at his wits end to know just what to do in the circumstances,

Part of the panorama from the Coronation Mountain station where the BC crew lost some of their field notes. V771_pd_15_76 and V771_pd_15_77, Whyte Museum of the Canadian Rockies

took the only common sense action possible by re-occupying them immediately. All the photographic views previously taken, were safe at the camp, but the transit readings for azimuths and orientations of the views, without which the views themselves were of little value, had to be done again.

On August 4, the party once more packed over the pass and were engaged until August 8 in reoccupying five of the six stations established in that vicinity.

On August 9, Wheeler and two packers (a new second packer, Charlie Plymensen, and Guy Thomas, who

SURVEYING THE 120TH MERIDIAN AND THE GREAT DIVIDE

was hired temporarily) left Field. When they arrived at Howse River two days later, Wheeler found no horse tracks going downstream so the men camped there. The next morning they travelled up Forbes Brook to the main camp, where he found a letter from Campbell describing the accident. Wheeler and Plymensen proceeded to the camp below Bush Pass. During the next two days, the crew finished surveying in the area. On August 15, they returned to the junction of Howse River and Conway Creek, where Wheeler found that a bear had torn part of the tent he had left there. The crew spent four days at this location surveying two stations while dealing with inclement weather and horses who had strayed away. One station was on nearby Mount David, named for the famous explorer David Thompson.

The crew then headed north down Howse River, a tributary of the North Saskatchewan River, and established a camp about 2.5 kilometres from the mouth of the creek flowing out of Glacier Lake. This creek had its origins on the east side of the Great Divide in a sizable valley. After a day of rain, the crew started on a fly camp up the valley above Glacier Lake on August 22. Wheeler noted that it was "surrounded by ice at the end coming from Lyell Icefield and icefield west of Mt. Forbes." After two days of rain, the crew resumed surveying on August 25. Wheeler wrote that the station was

> a very fine one and appears to be a peak of the Great Divide. Was much surprised to find the divide came around this way. Expected to have to go several miles back on the snowfield. Call the station Glacier Lake W. Excellent station with grand views all around. From peak divide runs northwest across snowfield to Mt. Lyell. Got back to camp after 7 pm—very satisfactory day. Left instruments cached on the moraine below our peak to be ready for tomorrow.

Rain prevented surveying until August 27.

> Made the ascent of our peak without much difficulty. Called the station Mt. Arctomy S [Greek word for marmot] on account of

the many whistlers in the alpine valley below it—saw 7 in one group and others around. Good station for the divide and for the valley of the lakes. Pretty cold for a while. Got back to camp just before dark.

Wheeler tried to survey another station on a small peak west of Mount Sullivan the next day.

> Had a lot of climbing to get to peak. When about 300 or 400 ft [90 to 120 metres] below summit saw it looked hopeful, so stopped and had our luncheon. Waited there trying to keep warm till 3 pm when gave up and returned to camp. Very cold with sprinkling of snow.

On August 30, Wheeler, Campbell and Cameron were able to complete the surveying: "Got two stations and some pretty good views but clouds hid western distance."

During the last day of August, Wheeler attempted to survey a station at the head of the Forbes snowfield.

> When we got to where instruments cached it had clouded up and soon began to rain. Got shelter under an overhanging rock. After a while at 10 we started out to cross snowfield but when had gone a little way clouds rolled over it enveloping us and travelling on it became dangerous, so turned back. Heavy rainstorm gave us a good wetting. Continued back to camp through rain and when we got in were really wet.

Wheeler also wrote: "Fell on the ice on the way home and hurt my back."

In his diary for September 1, Wheeler commented that his back was sore. The weather cleared in the morning, so he did a station nearby: "Left at 11:30, went up on high point of moraine between two glaciers—no trouble getting there from the ice. Took 8 views which should be very spectacular of the icefalls. Enjoyed the station although back bothered me quite a bit."

A member of the BC crew at the Bush Pass South station. V771_pd_15_92_001, Whyte Museum of the Canadian Rockies

With a full day of work on September 2, Wheeler was able to complete his surveying in this valley.

> Breakfast at 5, got away at 6 and made place where instruments were cached at 10 am. Body of clouds rolling up Glacier Lake valley and Bush Cr[eek] valley gave us a scare but they dissipated. Took us 2½ hours to cross snowfield—slow work as the new snow made travel among the crevasses dangerous. Reached arête of our peak— the watershed line—at 10:45 am. Started to climb arête. Climbing required care and was slow. Reached summit of Snow Pk. at about 1:30. Very fine station and quite warm for work. Took two photo stations and a round of azimuth. Built a large cairn. Left at 4:15. Had to hurry to get home before dark. Took an hour & a half for arête and crossed snowfield in much quicker time. Made camp by 9 pm. It was just about dark when crossed icefall at camp—at last successful—very fine sta[tion]. All pretty tired and back pretty lame all day.

One of the photographs from the Mount David station. Wheeler has marked the mountains visible in this picture. V771_pd_16_112, Whyte Museum of the Canadian Rockies

During the crew's return the next day, "on [a] hill near camp one horse rolled back and [we] had to unpack him to get him up." Wheeler also commented: "Back still very sore." While Wheeler rested, Campbell, Thompson and Cameron surveyed a station near camp on September 4.

Wheeler's final destination for the season was Thompson Pass, farther north along the Great Divide. To reach this location, the crew headed down Howse River to the North Saskatchewan. Then they proceeded up this river, following the general route of the current Icefields Parkway north of Saskatchewan River Crossing. Their campsite was "near an old Indian camp where there were a lot of teepee poles to borrow." Wheeler wrote that he "slept in open—did not put up my tent. Glorious night—a blaze of stars." On September 6, the crew spent "all day on gravel bars and mud flats of river valley—crossing and recrossing channels of same," and that night "camped at bend of Alexandra River." Once again, Wheeler slept in the open. The third day of travel took the men up the Alexandra, a large tributary of the North Saskatchewan that had its

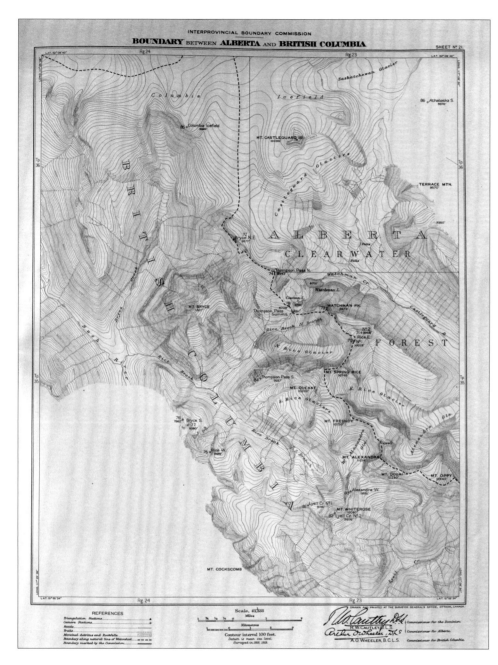

This map covers the terrain around the Thompson Pass area where Wheeler surveyed during the latter part of the field season. "Boundary between Alberta and British Columbia" map atlas, Sheet 21, Land Title and Survey Authority of British Columbia

had quite a hill and only a horse trail to follow. Passed a glorious little lake, then climbed another hill and camped beside another little beautiful lake."

Wheeler and his crew spent the next week surveying in the vicinity of Thompson Pass, which was located on the Great Divide. Fortunately they had good weather. On September 9, after a day of rest and some reconnaissance, Wheeler surveyed at a station from which he was able to see several of the stations he had established earlier in the season and also in 1917. The next day, the station they surveyed provided a good view to the north and the Columbia Icefield. In his diary, Wheeler wrote: "Climbing rotten scree, boulders and hard ice—had to spend an hour and a half cutting steps in hard ice.… Not very tired considering I have climbed for the last three days. Goat seen close above camp. Cook & Charlie hunted same but did not get any." September 11 was a "very fine day again," and Campbell and Wheeler each surveyed a station.

Then the crew went on a fly camp, crossing Thompson Pass, and travelling about 6 kilometres down the north branch of Rice Brook on the BC side of the Rocky Mountains. They cut a trail so the horses could be brought along. Although they located a good campsite, there was no feed for the horses, so they had to be tied up at night and taken back to the main camp the next day. During the next two days, Wheeler surveyed two stations from the west side of the divide, before returning on September 15 to the camp in Thompson Pass, where Wheeler occupied one additional station.

The crew then returned to their September 6 campsite on the Alexandra River, and on September 17, they climbed a peak directly west of the camp. "Very fine day and very good views, which made clear the north and west faces of Mt. Lyell. Station situated on the north side of the Alexandra River." Then they travelled about 10 kilometres down the Alexandra River and surveyed a station from there the next day. Wheeler wrote:

> Had horses brought in and with Campbell, Thompson and Cameron rode horses 9 miles [14 kilometres] up the valley to a suitable place to climb. Charlie with us—packed instruments.

headwaters along the divide. Wheeler wrote: "Trail lay for most part through bush. Good at first but as turn off to Thompson Pass was neared got very bad and took in spots to the gravel bars. On turning off to pass

Part of the panorama from the Thompson Pass North station. V771_pd_18_229 and V771_pd_18_230, Whyte Museum of the Canadian Rockies

Started to climb at 8:45 and arrived at top at 12 o'clock.… Called station Mt. Saskatchewan NE—good station and views would have been good but for haze & bush fire smoke which rolled up in clouds over Mts. Lyell and Alexandra. Took a sta[tion] from which we could see Observation Peak [a station near Bow Summit used in Wheeler's 1917 survey]. Got to horses after 6 and to camp just before dark.

Now, on September 20, with food running low, Wheeler and the crew were ready to make the return trip to Lake Louise. Wheeler noted that "he met a party of Indians" the first day, but provided no further detail. He also encountered guide Jimmy Simpson with a man on a hunting party: "Heard that war news was OK." By the next evening, the crew camped a few hours below Bow Summit. Wheeler, Campbell and Cameron attempted to survey at Observation Peak but were unsuccessful because of clouds. The crew camped that night at Bow Summit. Campbell and Thompson remained to survey Observation Peak while the rest of

the crew moved on to camp at Mosquito Creek. On the afternoon of September 24, they arrived at the Lake Louise train station. While the packers brought the horses to Banff over the next two days, the rest of the crew took the evening train. On September 27, Campbell sent Wheeler a telegram from Lake Louise informing him that he and Thompson had completed the 1918 field season with the survey from Observation Peak.

Many aspects of the 1918 field season were difficult for Wheeler and his crew. In describing the terrain of the Great Divide surveyed during that year, the Boundary Commission report commented: "From Howse Pass to Thompson Pass the course of the watershed, although in general direction northwesterly, is very erratic." The report also noted: "Probably the most striking feature of this section of the watershed is the numerous great bodies of ice and snow that are accumulated along the crests of the main divide range and the large number of wonderfully broken icefalls they send down in every direction." This section featured several peaks that are over 3000 metres in elevation. In a letter to Deville after the field season, Wheeler described how the difficult terrain affected the surveying.

Cautley's crew leaving Grande Prairie June 20, 1918. e011025438, Library and Archives Canada

The high mountain region surveyed last summer is the most difficult we have encountered, and particularly on the British Columbia side, where the only access is by going over the divide and travelling down the streams on the other side. As horses can only be taken over the Thompson Pass, the first accessible to them north of Howse Pass, all material had to be carried on our backs, making it extremely difficult to keep up food supplies and carry bedding, and entailing considerable hardship through an insufficient supply of both.

Besides being a difficult area to survey, Wheeler did not have a very experienced crew, particularly for climbing the mountains. This was especially apparent when he departed for the Alpine Club camp, leaving Campbell with two men who were still novices on the high terrain. The loss of the field book on Coronation Mountain necessitated several days of repeat surveying. There were several other mishaps during the summer, but fortunately none of them were too serious. The weather was inclement for much of the time. The crew lost several days of work, and when they surveyed, it was frequently difficult to take good photographs. A rainy summer kept the water level of the rivers high, and during the frequent crossing and recrossing of the gravel bars on the rivers, some of the camera cases got water in them and a few of the glass plate negatives were damaged. Wheeler wrote to Deville:

> I regret to say that Camera No. 243 and holders marked X got wet, owing to the

pack horse carrying them taking a swim when crossing the Saskatchewan. I examined the boxes containing the cameras and holders as soon as we unpacked and they did not appear to have got wet, but later I discovered that a little water had got in. By that time the damage was done.

All summer long it has been an incessant crossing of channels traversing the wide gravel flats of the Saskatchewan and its tributaries. The water was very high this year and one never quite knew when it would be a matter of swimming for the horses.

R.W. CAUTLEY

Cautley's 1917 field season finished at Yellowhead Pass, west of Jasper. He knew that he would be surveying the 120th meridian in 1918, so he left his horses to winter with the Otto brothers from Jasper. In February, they informed him that "owing to severe storms and the fact that the snow was badly crusted, the horses were in danger of starving." Cautley paid for six tons of hay and had it transported from Edmonton.

Like surveying the passes, delineating the 120th meridian was labour-intensive and slow work. Most of the meridian line was through thick forest, so the axemen were constantly busy. The skylight needed to be 1.8 metres wide, which meant that the boundary line had to be cut at least 3 metres wide at ground level, depending on the thickness of the timber and the size of the trees. The line had to be surveyed, distances measured, elevations read, and concrete monuments constructed at the stations. Packers and a cook were also part of the crew.

Cautley did not have an assistant surveyor in 1918. Like Wheeler, he had difficulty finding men to employ. Cautley, his crew and his horses left Edmonton on June 13, travelling to Grande Prairie on the Edmonton, Dunvegan & British Columbia Railway (ED & BC, nicknamed the "extremely dangerous and badly constructed"). In his Boundary Commission report, Cautley noted that "the road bed was in very bad condition and was the cause of various delays due to derailments and a wash-out on Smoky Hill." At Grande Prairie, Cautley

hired two more men. He and his crew left Grande Prairie, Alberta, on June 20 and arrived at Pouce Coupe, BC, on June 26, a distance of approximately 135 kilometres. The longitude pier was about 120 metres from the telegraph station.

Cautley wrote a memoir, and in it, he recalled that his crew had included a sixteen-year-old boy, a former soldier named John MacPherson, and a young lad that he had to train as a packer. More ominously, Cautley hired Fred Plouffe, who he soon found had a criminal record and was suspected of murdering two people.

When Cautley arrived at Pouce Coupe, he was met by a sergeant from the Royal Northwest Mounted Police. The night before the survey crew left Grande Prairie, two homesteaders had been murdered, and he wanted to know whether Cautley had any information about his crew members, particularly Plouffe. Three weeks later, a sergeant came to interrogate Plouffe. He decided not to arrest the man on suspicion of murder, but on the same day, Plouffe left the crew. Cautley wrote in his memoir: "I don't believe for a minute that he had anything to do with the murders, but I have never seen a worse exhibition of a bad conscience under police questioning." Less than a week after the first killings, four more people were murdered near Grande Prairie. The murders were never solved.

The longitude pier was 8.3 kilometres west of the 120th meridian, so it was imperative to measure the exact distance to this location. The Boundary Commission report stated that

the connecting survey called for extreme care since any error in its measurement would have resulted in the entire boundary being out of place by the full amount of such error. The connection was made along the north boundary of Township 77 and chained twice by the transit and tripod method, after the chain had been very carefully compared with standard tapes.

As soon as this survey had been made, concrete Monument 78-1 was built 3.079 chains [61.6 metres] north of the north boundary of Township 77 and formed the initial point from which the survey of the

Pouce Coupe in 1918—restaurant, trading post, Canadian Bank of Commerce, fur buyer's office. e011205440, Library and Archives Canada

120th meridian was subsequently carried north and south. (Along the 120th meridian, the stations were numbered based on the township in which they were located, with the township numbers increasing to the north. Within the township, stations were numbered [1, 2, 3, etc.] with the numbers also increasing to the north. Each township was 9.6 kilometres [6 miles] in length.)

Along the 120th meridian, a large portion of the line through the Peace River district was the eastern boundary of the Peace River Block, part of the land grant to the Canadian Pacific Railway for construction of the transcontinental railway. Monument 78-1 was about 10 kilometres from the southern boundary of the Peace River Block. At the request of J.E. Umbach, the surveyor general of British Columbia, the Boundary Commission directed Cautley to survey south. After reaching the southern boundary of the block, he was to continue for 32 kilometres. This land was owned by the BC government and there were applications for land in the vicinity of the boundary.

On July 30, from a camp on the 120th meridian, Cautley wrote to Deville, reporting that he had sur-

veyed from the longitude pier to the meridian and 6.4 kilometres south "through thick bush" to the intersection with the Pouce Coupe River. His second packer, Plouffe, quit on July 22, and "my head packer has ridden off 60 miles [96 kilometres] to a doctor with an abscess in his throat which prevented him from eating." He had been unable to get another packer. Cautley, like Wheeler, had many days with heavy rain.

In his memoir, Cautley wrote a few observations on the effects of the war.

> Going up through the splendid farming country from Pouce Coupe to Rolla, it was pathetic to see all the young homesteaders' cabins: windows and doors boarded up; weeds growing rankly on their earth roofs and all over the untracked yards, and their few head of stock sold or farmed out to a neighbour "for the duration."

He also wrote that during the summer he received a telegram telling him that John MacPherson's

Longitude pier established by F.A. McDiarmid at Pouce Coupe in 1917. e011205435, Library and Archives Canada

son had been killed in the war, and he had to inform his monument-builder of this news.

In his annual report, Cautley described the difficulties he encountered during the 1918 surveying.

> The surveying of the 120th meridian involves the same kind of work as surveying a principal meridian or base line of the Dominion lands system, and because it is an interprovincial boundary, it involves additional labour in cutting a more than usually wide line through timber, and in constructing monuments of a much more enduring character than are called for on base line surveys.

He noted that normally there are at least fifteen people on these crews, but

> during the past summer I had a maximum outfit of seven men besides myself.... As the greater part of the line surveyed is in

thickly wooded country, it meant that for days at a time the whole available party, including myself had to chop line, or were engaged locating and cutting horse trail, moving camp, or some other labour. Just as soon as the line was cut far enough ahead to carry forward the survey, it meant that three of the above six men were taken from the line cutting to do the chaining and the second packer had to go off after gravel for the monuments, so that two men only were left to carry on a line.

Cautley's crew made slow progress south, and by the close of the field season at the end of September, he was about 10 kilometres short of his objective. The crew arrived in Grande Prairie on October 10, and Cautley decided to sell his pack train, because he was concerned about arrangements for wintering the horses in the Peace River country.

Bear trap on shore of Swan Lake. e011205441, Library and Archives Canada

In addition to the difficulties Cautley and Wheeler encountered during the 1918 survey season, the establishment of the 120th meridian created a political controversy that extended beyond the work of the two boundary commissioners. In 1911, during the initial survey of the Peace River Block, British Columbia surveyors used their determination of the 120th meridian as the basis for establishing its eastern boundary. McDiarmid's observations and Cautley's surveying found

Left: Crew member using level rod for measuring elevation at Monument 77-2. e011205437, Library and Archives Canada

"Approaching Saskatoon Lake from west." e011205444, Library and Archives Canada

that the actual boundary was about 350 metres to the east, adding a new strip of land to British Columbia.

After completing the fieldwork for 1918, the BC government hired A.J. Campbell to go to the Peace River district. In the area where Cautley had surveyed, Campbell was to survey this strip into lots. In his government report, Campbell described the unfortunate results.

> We left Grande Prairie with an incomplete party in the hopes of being able to obtain men from among the settlers around Swan Lake. Men were secured and the work was proceeding with the greatest possible dispatch, when a member of the party was taken ill with the Spanish influenza. This dread disease spread though the party until, finally, every member except one was either down with it or convalescing from it. I was thus compelled to cease operations with the work allotted to me in an incomplete state.

The British Columbia and federal governments had to decide how to readjust the Peace River Block so that the new strip of land was included but the acreage remained the same. To resurvey the western boundary would have been the simplest solution, but time and expense made this option impractical. Over the next few years, the question of which parcel of land would be removed from the block was discussed. During his thirteen years as surveyor general of British Columbia, the disputatious J.E. Umbach became embroiled in several controversies with the Dominion government and this was one of these episodes.

"The launch of the *Fortress Queen*," the raft that the BC crew used to travel around Fortress Lake. PA10724, Provincial Archives of Alberta

1919

In the fall of 1918, Wheeler returned to his home in Sidney, on Vancouver Island, while Cautley went to his place in the Edmonton area. During the next six months, the men would work on their survey calculations, produce maps and label photographs, and attend to other administrative details. Then they would begin to make plans and prepare for the upcoming field season. During the first five years of the survey, while the two crews covered the same area, their data and maps were connected, so there was considerable interaction between Cautley and Wheeler. Now, however, the two men worked separately. Wheeler continued his mapping work along the Great Divide, while Cautley began mapping the 120th meridian. He was surveying only the meridian and not the surrounding country, and the terrain was relatively flat during the first years of this part of the survey. As a result, Cautley's maps were much less complex, and there were only a few places like lakes or river crossings where he needed to add more detail.

World War I ended in November 1918, and by the spring of 1919, both the federal and provincial governments were beginning to spend more revenue on projects that had been cancelled or curtailed. After four years of operating on a minimal budget, the Boundary Commission looked forward to accelerating the pace of the survey. Wheeler wanted to have sufficient, experienced personnel to run two small survey crews, while Cautley needed more personnel for his labour-intensive survey of the 120th meridian.

A.O. WHEELER

Wheeler eagerly awaited the return of Ralph Rink and A.S. "Spike" Thomson, two members of his 1916 and 1917 survey crews who had served in World War I in 1918. On March 22, 1919, Wheeler wrote to the BC minister of lands and told him: "I badly need Mr. Thomson's services this coming season, owing to his training and skill, and his presence is essential to the survey in the very difficult section of country that lies before us." He also described Rink's importance: "Ralph Rink was my head packer and I need him badly again. He is a skilled man, trained with special aptness for the work." Wheeler exhorted the minister:

> These two men are necessary to enable me to carry on the work of the Boundary Survey this year with full success. As there does not appear to be any possibility of my obtaining their return through my personal effort, I am appealing to you in the hope that you will take the matter up in the interests of British Columbia.

On May 17, Wheeler sent a telegram to "Air Mechanic, A.S. Thomson, RAF, Passenger, S.S. *Saxonia*. Report to me at Banff ASAP." In addition, the famous mountain guide Conrad Kain, who had been a member of Wheeler's crew in 1913 and for part of 1914, joined the crew. Kain hoped to climb some of the high peaks of the Rockies that were in the area to be surveyed in 1919. He was also planning to spend the upcoming winter trapping in the vicinity, so the survey work would provide Kain with an opportunity to become familiar with the terrain and begin preparations. A.J. Campbell returned as Wheeler's assistant, along with Tom Martin, who had been the cook for the previous two years. One new person was Walter Nevler, a young man from Calgary. Although Wheeler had an experienced group of men for his 1919 surveying, this did not prevent some mishaps from occurring.

Crossing Bow Summit. 89.03.129, Jasper Yellowhead Museum & Archives

Spike Thomson had kept a diary during his work in 1916 and 1917 to which he later added more detailed notes, and he did the same for his time on the 1919 and 1920 crews. On June 5, he arrived in Banff, where he met Wheeler and made arrangements for the coming season. That evening he left for Victoria, where he spent two days before travelling to Vancouver with A.J. Campbell and his wife. By June 12, Thomson was back in Banff, and the following day he rejoined the boundary survey as a draftsman for office work and second assistant on the field crew. He met Arthur Hughes, who had been hired as a second packer and who would work on Wheeler's crew for several seasons. That evening the two men played billiards and went to a movie, two of Thomson's favourite activities when he was in a town. During the next week, Thomson tested the phototopographic cameras for accuracy, got supplies ready, helped finish the 1918 maps and packed up. Ralph Rink and Conrad Kain arrived. The day before departing, they spent the afternoon at

Brewster's corral. According to Thomson, one of Brewster's employees and Wheeler "had a run-in over horses, apparently they had been using a string set aside for the Boundary Survey. AOW became very hostile."

On June 20, the pack train left for Lake Louise. The survey crew left by train the next morning and met the packers when they arrived in the late afternoon. After a Sunday in camp making the final preparations, Wheeler arrived on June 23, and the crew departed late that morning. They followed the general route of the current Icefields Parkway north to the area where they had finished surveying in 1918. The crew crossed Bow Summit on the second afternoon and, on the evening of June 25, camped at the North Saskatchewan River crossing. In his diary, Wheeler noted that the river did not seem very high. They "forded horses without difficulty" the next morning and then proceeded up the North Saskatchewan valley. Thomson wrote that they camped that evening about 3 kilometres below their site in 1917 and that a "party of Stoneys camped

Surveying the 120th Meridian and the Great Divide

a little below us. Chief came over and had supper." In a later note, he added: "Samson Beaver—good-natured, jovial, could converse well in English, muscular, wore moosehide trousers." Before leaving the next morning, Samson "traded big chunk of sheep meat for sugar, tea, flour." That day Wheeler and his crew travelled up the Alexandra River and camped on the flats below Castleguard Falls. By the evening of June 28, they had reached the Thompson Pass area and made camp in the same location that they had used in 1918. Wheeler noted: "Very little for horses to eat. Snow not long off of pass and spring only just commencing."

After Sunday in camp on June 29, Rink departed for Lake Louise with eight horses to pick up more supplies. The survey crew left on a fly camp across Thompson Pass to the site of their 1918 camp near the junction of the north and south branches of Rice Brook. During the winter, while mapping the 1918 survey, Wheeler had found that he had incomplete data for the western side of the Great Divide in this area. He also hoped to get a station down in the Bush River valley.

During the first three days of July, Wheeler and his crew located and cut a trail up the south branch of Rice Brook to timberline. On July 4, Martin, Hughes and the remaining horses came to the fly camp. Wheeler decided to cut a trail down the main branch of Rice Brook to the Bush River, because he "wanted to occupy a station on far side to gain further control of western limits of Columbia Icefield." Thomson wrote in his diary:

> Not long before ran into trouble—below junction of north and south branch valley narrows, runs into canyons. Trail became slow and very difficult. About 11:30 had lunch beside first big canyon. Meanwhile AOW had gone on ahead to try to locate a possible route. Hit heavy timber, windfall, alder slides.

Thomson recorded that Kain and Hughes had shot a goat and they had goat soup for supper. Rain kept the men in camp until the afternoon of July 7, when they resumed cutting a trail to the Bush River. The steep canyons along Rice Brook forced the men to climb until they reached a ridge, where they saw that they were about 900 metres above Bush River on precipitous slopes and could not get down to the valley: "We stood silently gazing out over some of the most rugged country we had encountered in the survey." Thomson noted that they had not yet occupied a survey station.

On July 8, Wheeler and the crew headed up the south branch of Rice Brook. Wheeler, Kain and Nevler cut trail, while Hughes, Campbell, Thomson and Martin brought the horses. Thomson described the accident that occurred.

> We got as far as the big hill—climbing up under the big rock outcrop, one of the horses up front (Pete) rolled on a steep turn on the trail. This threw two other horses off balance & soon the whole string became restless. Five horses altogether got into the panic. The first one by now was out of control & began to roll down the steep hill end for end—I got a bleacher view of the whole pandemonium as I was below bringing up the end of the string with AJ. They were all coming down the hill in confusion. It was fortunate for AJ & I [that we] were standing close to a big fir tree, as by now the first horse, Pete, a buckskin, was catapulting through the air directly to where we were standing. We made the shelter of the tree on time. The poor beast then collided with the stump of a tree that had been cut 3 or 4 feet [0.9 to 1.2 metres] from the ground—scattering the pack high wide & handsome. By now four other packs had been shed in all directions over the hillside above us.

Pete was severely injured and the men spent the rest of the day rounding up debris, before reaching camp at supper. Thomson continued:

> In the evening we stretched Pete (who had been led into camp) on the ground between two suitable trees, anchored his feet and hind legs with cinch ropes. Surgeons extraordinaire—Campbell, Hughes

Columbia Icefield and Mount Castleguard (centre right). v771_pd_20_46 and v771_pd_20_47, Whyte Museum of the Canadian Rockies

and Thomson—went to work to sew up a nasty gash in Pete's stomach—a 6" by 8" [15 by 20 centimetres] tear in the skin. Heavy black thread, hot water, darning needle, antiseptic. Had the poor animal up on his feet in an hour.

The next day the crew moved back up the creek—without the horses. Their camp was in the subalpine under the snow below the tongues of glaciers from Mount Alexandra and Mount Whiterose. On July 10 and 11, the crew surveyed three stations at the head of the creek and obtained the information needed to map the area. They then returned to the camp near the junction of the two forks of Rice Brook. After a day of rain, the crew surveyed one more station on a peak that overlooked the north branch of Rice Brook and Thompson Pass. Before leaving the area, Wheeler made one more attempt to reach the Bush River valley on July 14. Wheeler wrote in his diary: "Got away at 10 am and travelled down canyon of main Rice Brook till 12:15 when had lunch. Soon after struck a snag and could not go further without such difficulty that to go on meant failure. So decided to turn back and give the expedition up." During this outing, Thomson slipped on some moss and hit his head on a rock. He experi-

enced dizzy spells and had a nosebleed, so the crew laid him down on his back and packed snow around his head and neck until the bleeding stopped.

The crew spent July 15 resting and preparing to leave. Pete died from the injuries he had suffered a week earlier. The next day, the crew moved back to their camp of June 27 along the Castleguard River. Sleet and fresh snow at timberline kept the men in camp the next day. On July 18, the men cut a trail up to the subalpine south of the Columbia Icefield where the Saskatchewan Glacier flows from it. With Kain as guide, Wheeler, Campbell and Thomson travelled out onto the most westerly part of Castleguard Glacier on July 19. They surveyed from a station that provided views of the Columbia Icefield as well as some of the terrain at the head of Bush River. Thomson wrote: "Glorious panorama from station. Fine peak. AOW saw Sir Sandford in Selkirks. Icefield heavily crevassed, compelled to rope, several snow bridges." After finishing this station, Wheeler departed with Rink to attend the Alpine Club camp in Yoho Valley.

From July 20 to 22, the crew surveyed three stations along the southern border of the Columbia Icefield. One of these was Mount Castleguard. Thomson noted that this was a key location in their triangulation surveying, and they would see this station in September from the

Part of the panorama from the Athabasca South station. Mount Castleguard is the snowy peak in the left centre, while Mount Columbia is in the centre back. v771_pd_20 50 and v771_pd_21_51, Whyte Museum of the Canadian Rockies

head of the Athabasca. He wrote: "The panoramic view from the summit of Mt. Castleguard has been described by several who made the climb.… Unfortunately, later in the fall at the Sidney office, what would have been one of our finest panoramas was not to be. We found to our dismay that during the process of developing at Ottawa the whole set of exposures were spoiled."

The men spent July 23 resting, except for Kain. Thomson wrote: "Conrad down to camp below falls to work on cabin. He has decided to spend the winter trapping." On July 24, the crew attempted to climb Snow Dome. This high peak of 3456 metres occupied a prominent location near both the Columbia and Clemenceau icefields. At Snow Dome, the divide changes direction from north to south-southwest. When the crew was about 1.5 kilometres from the summit, the weather began to deteriorate and they had to return. Thomson described a weather phenomenon related to the icefields: "During the summer months, especially on warm days, huge billowy clouds form over the Icefield, towering high over the summits, gradually to settle down & cut off all visibility, clearing in the cool of evening—referred to as steaming."

To continue surveying along the Great Divide, Campbell and the BC crew had to retrace their route to the North Saskatchewan River and then proceed north

for several kilometres, approximately following the route of the Icefields Parkway. The next day, as they were packing and preparing to move camp, Campbell had an accident. Thomson wrote:

> Bagley, a restive beast who would seldom stand once the load was on, began to act up while they were cinching the "diamond." Suddenly he reared up wrenching the halter, of which I had hold, out of my hand & began to take off. AJ, who had been making a cinch over the beast's rear, had one foot in a loop of the cinch rope on the ground. He lost footing, but the coil tightened around his ankle before he could save himself.

Campbell injured his knee, which limited his work during the next three weeks. "Horse was reloaded with an extra 100 lbs [45 kilograms], including a wet tent, to keep him cool."

Two days of travel brought the crew down to the North Saskatchewan River. Thomson noted that Campbell's knee was "badly swollen." On July 27, the crew continued up the river, climbed the big hill leading out of the valley and made it to Camp Parker,

Setting up at Camp Parker. 89.03.142, Jasper Yellowhead Museum & Archives

Tree tablet. PA64-3A, Jasper Yellowhead Museum & Archives

the main campsite in the area. Thomson described the difficult day of travelling: "A.J. indisposed. We had to keep him in the saddle. I doubled up with him at the fords." They rode Bagley across the river flats below the junction of Nigel Creek. On the far side, the channel deepened and the horse couldn't get out, so Thomson jumped into the ice-cold water and grabbed the horse's tail. The current swung Thomson into the bank, and then Bagley was able to make it out of the river. They reached the foot of the big hill at 3:00 pm, and arrived at Camp Parker at 4:00. "By the time we reached the campsite—under a foot [30 centimetres] of snow—the weather had turned bitterly cold. Snow was falling. The ground was frozen." Thomson described Campbell's situation:

> Before pitching the tents, our first concern was for AJ, who by this time was just about frozen, unable to get out of the saddle and walk. The knee by now was badly swollen. We unloaded the horses, made a bed beside a tree with some of the horse blankets and laid him against a tree in sitting position and covered him with remaining blankets. In this manner kept him warm until we could get a fire going.

On July 29, after a day of rest, the crew surveyed from a nearby station on Nigel Ridge. Thomson was temporarily in charge of the surveying, while Campbell, the cook and the packers stayed in camp. The next day they surveyed from an "outlier of Mount Athabasca, overlooking terminal moraine of Saskatchewan Glacier. A low station to help strengthen our network around the southern extremity of the Icefield."

The crew moved camp to Wilcox Pass on the last day of July. In 1919, the trail went above the present highway across the pass to Tangle Creek, and then dropped steeply to the Sunwapta River. The campsite that the BC crew had used in 1917 was occupied by Claussen Otto (one of the Otto brothers, operators of a guide business in Jasper), who had a group of tourists with him, so they set up their camp about 450 metres farther up the trail.

The survey crew spent a week at this camp. August 1 was a day of rest. Thomson noted that they "made tree tablet, names of crew members and decoration." Campbell spent the time cleaning and repairing a transit, using Thomson's bed as a bench. When Thomson returned to the tent, he flopped down on his bed, scattering the transit pieces. "Poor AJ remained speechless—a distant look in his eyes." Eventually they retrieved all but one small screw.

On August 2, the crew surveyed a nearby station that provided an excellent view of the area. Poor weather then kept the men in camp for a few days. Otto and the tourists also remained at their camp. Claussen came to visit a few times, and one of the tourists brought some magazines for the crew. On August 5, the crew tried to survey from Nigel Peak, but turned back as the weather deteriorated. Two days later, the men returned and reached the peak, but the weather deteriorated again, so they were only successful in doing the surveying part of the station. Thomson noted that it was a "difficult descent down snow and ice."

The survey crew moved camp to the head of Tangle Creek on August 8. The same day, Wheeler began the trip to rejoin the crew. After completion of the Alpine Club camp, he had returned to Banff for a few days and from there took the train to Lake Louise. Wheeler wrote: "Ralph met train & had horses ready.... Went out on Bow Valley trail for 7 miles [11 kilometres] where Frank Wallis had lunch ready for us. Packed up and pushed on up the valley—Frank with us. Camped a short distance beyond Mosquito Creek. Rained hard during night." (Wallis had been assistant packer to Rink on Wheeler's 1917 crew and had travelled the same route. Rink must have hired him for a few days.) The next day Wallis accompanied Rink and Wheeler. From their camp that night, Wallis returned to Lake Louise carrying the contents of a telegraph to send to Wheeler's wife.

Meanwhile, the crew surveyed at a couple of stations on a nearby ridge that provided an excellent view of the Sunwapta River drainage; this river was a tributary of the Athabasca River. Campbell came for a walk. They surveyed one more station in the area the next day, but Campbell did not accompany the crew.

Crossing Wilcox Pass. 89.03.185, Jasper Yellowhead Museum & Archives

August 11 was spent in camp. "AJ nursing his knee," Thomson wrote.

Wheeler and Rink continued north to join the crew. Travelling down the Mistaya River, the two men passed Jack Otto with a tourist. They forded the North Saskatchewan without any difficulty and "travelled up Saskatchewan valley to camp by Indian teepees where arrived at 4:30 pm. Camped for night (1918 camp)." On August 12, the two men camped at Wilcox Pass where they found a note from Campbell. That same day, Campbell's crew had travelled down to the Sunwapta River.

On August 13, Wheeler arrived at the Sunwapta River camp in early afternoon where he found "Campbell confined to camp with his knee." Meanwhile, Thomson, Kain and Nevler were attempting to reach the summit of Mount Sunwapta (3315 metres) directly above camp. In his diary, Thomson wrote: "Long drag up shale slope 5000 feet [1525 metres]. About 11:30 became doubtful weather, returned to timberline and lit fire. Wait until 3 when storm moved in."

From their location in the Sunwapta valley, the Alberta-BC boundary turns west for several kilometres. Wheeler decided that he wanted to survey stations close to the boundary again, so the crew headed north to Fortress Lake on August 14. They travelled down

Campbell (left) and Thomson on Chaba Glacier. A10725, Provincial Archives of Alberta

damage done except to plates 13–24." The next day four horses were missing, and it took all morning to locate them. They continued down the Sunwapta for four hours and stopped at the campsite Cautley had used in 1917. Once again they experienced difficulty. Wheeler wrote: "Same horse had again gone into the river and again wet the transit. Camera & plates again escaped. As the pack was behind decided to camp.... Made only a few miles.... Dried out transit as much as I could." On August 16, the crew reached the Athabasca River and started following it upstream. Wheeler wrote in his diary: "Slept out last night & enjoyed it. Beautiful night, moonlight, & starlight, very warm.... No accidents today."

The next morning the men reached the junction of the Chaba River, a tributary of the Athabasca. After fording both rivers, the crew followed a trail towards Fortress Lake, intending to reach the east end. Instead, the trail brought them to a location along the lake a few kilometres away. Thomson wrote: "When they got to the lake Buck and Bagley went in for a dip—had AOW and A.J.'s bedding." Since they would have to move the next day, the crew did not pitch any tents, and they tied the horses. On August 18, the crew had breakfast at 5:30 am, packed up and left early. They went back to the Chaba and found another trail that brought the crew to the east end of the lake.

Fortress Lake is about 10 kilometres in length and flanked by high mountains. It lies in British Columbia, with the Great Divide passing close by the eastern end of the lake. Wheeler needed transportation to travel around this body of water. The Otto brothers were supposed to have a canoe stored at the east end of the lake, but an extensive search proved fruitless. The men spent August 19 constructing a raft. Thomson described the activities. After cutting several logs near camp, "we build the 'Fortress Queen'—4 oars, mast, stern bracket for steerage sweep, 10 to 12 poles, 15 feet [4.5 metres] long." The raft was completed in the late afternoon and then the crew "launched it amid loud cheers, then brought it over to the mouth of a small creek just south of camp. After supper a raffle was held of 1 cigar for the most appropriate name. I pulled the winning number for Arthur who suggested the 'Fortress Queen.' The christening was now at hand."

the Sunwapta and camped near the mouth of Poboktan Creek at the same site they had used in 1917. Wheeler described conditions that day: "Bad travelling along gravel flats and by side of Sunwapta River. Mud & bog, windfall etc. Horses went swimming and instruments (transit) got wet—fortunately no great

Surveying the 120th Meridian and the Great Divide

After surveying at a station near camp, the crew left on August 21 for a fly camp down the lake. Wheeler found that the "raft made a good deal of water and we got wet and chanced getting the stuff wet." The crew "put into a little cove and cut two more logs which Conrad lashed to the raft—one on each side. This helped a lot. Campbell also put a buffer board on front. Had lunch here and made tea. Walter had slipped in and was quite wet—lent him an undershirt." The men camped that evening at an old campground on the north side of the lake, a few kilometres from the west end. The crew spent the next day surveying from a station near camp, then on August 23 took the *Fortress Queen* to the west end of the lake. Thomson described the location: "Pitched tents at an old campground where we found teepee poles—evidence of Otto Brothers we surmised. Cabin a short distance away on grassy flat on N[orth] side Wood River flowing out of lake."

Unfortunately, smoke from forest fires moved into the valley and impeded surveying. After two days, the crew moved to the mouth of Chisel Creek where they set up camp. Wheeler wrote: "At night pall of smoke heavier than ever—absolutely prohibitive." The next day, his diary entry said: "Smoke still thick—no chance of climbing." On August 28, the smoke was replaced by showers and clouds.

Wheeler, Thomson and Nevler, guided by Kain, were able to resume surveying on August 29, occupying two stations from a high peak on the south side of Fortress Pass. The four men reached the peak after seven hours, and they had to rope up three times. Then it took them almost five hours to complete the surveying. "While taking azimuths smoke again came up from S[outh], got very thick." The men started back to camp around 6:30 pm but didn't make it before nightfall, "so we lighted a fire and spent the night on a gravel bar," Wheeler wrote. "It was a fine starlit night and the fire generally good, so we did not have too bad a night and slept some." As soon as there was daylight, the crew returned to camp, had breakfast and spent the morning sleeping.

Wheeler wrote that on the last day of August it "rained all day—snowed on the upper heights. Wind blowing up the lake.… Spent most of day under blanket." Clouds and rain precluded surveying the next day, so Wheeler sent the crew back to the main camp on the lake for more food.

On September 2, the crew surveyed at one more location in the area, about a fifteen-minute raft trip from camp. Wheeler described the busy day.

> Climbed to one peak—Fortress Lake Centre—arriving there shortly after 9:30 am—easy climb. Took views at Camera Station 1—dull. Immediately after light got good as is generally the case. Got better views later. Occupied two stations and took azimuths at cairn to locate them. Got through by 3:30. Got back to camp at 5:15 pm. Campbell had supper ready. Struck camp and started for main camp soon after 7 pm. Light head wind. Arrived at camp after 9 pm. Dark. Had to wade ashore with load. Cook gone to bed. Got to our tents as best we could in the dark & turned in. Main camp more soggy than ever after the rain.

The next day the crew got ready to survey up the Chaba River. Before departing, Wheeler and Thomson surveyed at the target that Campbell had set along Fortress Lake two weeks earlier. Their campsite that night was along the river near the Chaba Glacier.

They attempted to climb Chaba Peak on September 4, but after a few hours, it began to rain, so the men returned to camp. They tried again the next day, with the same results. Kain spotted a mountain goat. Wheeler wrote: "Conrad shot the goat—and Ralph & Arthur went up and helped him bring it in. Very necessary in camp as meat short." Rain kept the crew in camp for the next two days. During that time, they brought back the rest of the goat meat. The third attempt on Chaba Peak on September 8, by Wheeler, Thomson and Kain, was successful. The crew climbed by way of Chaba Glacier and had to rope up in a few places. The actual surveying was done below the summit, and the men got back just before dark. At the same time, Campbell and two men surveyed at a station on another mountain.

On September 9, the men moved up the Chaba River to the junction of the south and west branches

Part of the panorama of the Chaba Icefields from the Sundial station. In the background on the right is Tsar Mountain where Wheeler made the mapping error described later in the book. v771_pd_23_197 and v771_pd_23_198, Whyte Museum of the Canadian Rockies

of the river, and they surveyed a nearby station the next day. Then they started down the river and, after a day of rain, surveyed another station on September 13. When the crew reached the Athabasca River on September 14, Hughes left for Jasper to pick up supplies, while the rest of the men proceeded up the river for a few hours. Wheeler wrote: "Opposite this camp is a good peak to climb with an easy access." From there, Wheeler, Kain and Thomson climbed a northerly outlier of Mount Alberta, a station that took six hours to reach. While Wheeler and Thomson were surveying, Kain went on to a peak farther away. The men returned as it was getting dark. Wheeler wrote that he was "very tired—in fact exhausted."

The crew continued up the Athabasca to a location near Mount Columbia on September 16. "Ralph got a bad wetting in the morning when after the horses. They swam across and he had to follow on Rocky barebacked," Wheeler wrote in his diary. The next day Campbell and Kain surveyed from one peak, while

Wheeler, Thomson and Nevler occupied a station on another one. Thomson had an unusual experience while he was surveying at the edge of a steep wall with a 150-metre drop: "While taking readings almost blown off balance by sudden gust of wind. Golden eagle had swooped down for strike, threw me against transit."

On September 18, the crew returned to their camp of September 14. They remained there for a day before moving down to the junction of the Athabasca, where they met Hughes again. With clear weather on September 21, Wheeler and Nevler surveyed one station, while Campbell, Thomson and Kain went to the peak north of Mount Alberta that Kain had previously climbed. Since they were surveying for triangulation and not taking photographs, it took only 1¼ hours to finish the work. The next day the packers moved camp back to their location on August 26. Meanwhile, Campbell and a few men surveyed one more station, and then reached the new camp at dark.

SURVEYING THE 120TH MERIDIAN AND THE GREAT DIVIDE

From Blackfriars station looking towards Mount Columbia (back right). v771_pd_24_209, Whyte Museum of the Canadian Rockies

The crew was now ready to begin their return journey to Lake Louise. When they arrived at their camp of August 13 where Tangle Creek joins the Sunwapta, they stopped for a day to climb Sunwapta Peak, which they had been unsuccessful in surveying at that time. Wheeler described the day:

> Got away at 6 am. Soon clouded over and got very cold—at 10 am got a shelter out of the wind and waited developments. Light snow in squalls. Had a smoke & our lunches. At 12:30 started for the peak where we arrived at 1:30—5 hours actual climbing. Campbell set up first and got ready to work—clouds lifted and let us get azimuths required and some views. Then storms & clouds passed over again—but it was sufficient to finish the work.

While Wheeler and Campbell were surveying, Thomson began building the cairn, which also served as a windbreak. Thomson wrote: "So cold was the wind that to step out of the covey, level up the instrument was all AJ could do to read one angle, return to covey & huddle together to keep warm."

It snowed that night, and Wheeler realized that the crew had been fortunate to complete their surveying. Snow was still falling the next day, but the men needed to return, so they continued south, crossing Wilcox Pass and camping near the location where they had spent the first week of August. The next day they reached the North Saskatchewan River and were out of the snow.

On September 28, Wheeler's crew reached the North Saskatchewan River crossing and camped by the ford. There they met one of Brewster's packers bringing Kain's supplies for his winter trapping in the area. Wheeler paid his remaining wages, and Kain departed with the packer the next morning. Two more days of travel brought the crew to Lake Louise in the late afternoon. Wheeler noted that there was "plenty of grub for breakfast, but lunches shy." Thomson described the

Mount Columbia (back right) from Windsor Castle. v771_pd_25_274, Whyte Museum of the Canadian Rockies

J.T. Carthew. e011205471, Library and Archives Canada

scene: "At Lake Louise, taking packs off horses, sorting personal gear westbound train pulled in and became center of attention, much photographed. Pretty rough looking bunch, dirty and tattered clothes, perspiration from man and horses."

It had been a challenging season for the BC crew. There was the death of a horse and the serious knee injury to Campbell. The terrain was rugged and remote with many glaciers and icefields to navigate. Most of their work was from the Alberta side, since it was difficult to access the western slopes of the Rocky Mountains. This contributed to the most serious mapping error of the entire boundary survey, which occurred in 1920.

Johnson farm. e011205474, Library and Archives Canada

R.W. CAUTLEY

After considerable discussion among the three survey-or generals and the boundary commissioners, a decision was made that Cautley would begin by finishing the 120th meridian line south for 10 kilometres to the location requested by the BC government in 1918. Then his crew would travel back to Pouce Coupe and survey the meridian line north through the main agricultural area of the Peace River district. Cautley had sixteen men (including himself) and twenty-four pack horses.

According to regulations, Cautley was supposed to hire men returning from the war if they were suitable for the work. In his memoir, he described the crew: "All my men, with the exception of one boy and my head packer were returned men, but all were woodsmen, some of whom had been out with me before the war. I never had a better outfit in my life, or one that was so pleasant to deal with." At the beginning of the sea-son, on their way to where they would begin work, the crew stopped by a lake to camp one night, and the men decided to go swimming. Cautley was "staggered" by what he saw: "Every man had been wounded, except one, and their wounds had healed so recently that they still looked angry. The one exception, Ernest Brown, a fine young fellow who had been out with me before, had been at the front for all the war years without being wounded, but the psychological strain had broken his health, and he died within eighteen months." Cautley's assistant was J.T. Carthew, the younger brother of W.M. Carthew, who had been Cautley's assistant in 1914 and who had died during the war.

In the late spring of 1919, the Winnipeg General Strike occurred. There was also labour unrest at other locations, including the rail line between Edmonton and the Peace River region. This delayed the arrival of the Alberta crew. Cautley divided his party into two groups. One left Edmonton on June 5, and the second left four days later. The entire crew departed from

1919

In Cautley's personal photo album he drew the location of the 120th meridian line across the Peace River valley. j-01174, Royal BC Museum and Archives

Grande Prairie on June 14 and reached the end of the 1918 line on June 19.

The first two weeks were spent extending the 120th meridian south, and then it took two days to return to the initial monument that had been established on the meridian in 1918.

From July 9 to October 1, Cautley's crew surveyed the meridian 77 kilometres north, constructed fifty concrete monuments and connected their survey line with several survey posts in the vicinity. In addition, Carthew ran the levels for the line surveyed.

In Township 78, at the start of the survey northward, Cautley saw the effect that the establishment of the 120th meridian had for some of the settlers. The Boundary Commission report stated:

> Township 78 has been subdivided and, being within the Peace River Block, which is surveyed right across the Boundary as a normal extension of the Dominion Lands system, it follows that the Boundary cuts across the various farms in its path, so that these farmers find parts of their farms in the territorial administration of Alberta and parts in that of British Columbia.

During September, Cautley's crew surveyed the 120th meridian line across the Peace River, the steepest part of the line north of Pouce Coupe. Later that month, near the end of the field season, Cautley had an accident. In his memoir, he wrote that

> while rounding up stray horses, I got a bad throw in the saddle which set up such a bad attack of my ancient enemy lumbago that I was unable to turn over in bed. On the 1st October my assistant finished the season's work, and it was necessary to get down to Peace River in order to supervise the building of our homeward bound raft. I was still out of business, but two of the men hoisted me onto my saddle horse, and I rode fourteen miles [23 kilometres] down to the river over our rough survey trail. On arrival, I was lifted off the horse and helped to bed.

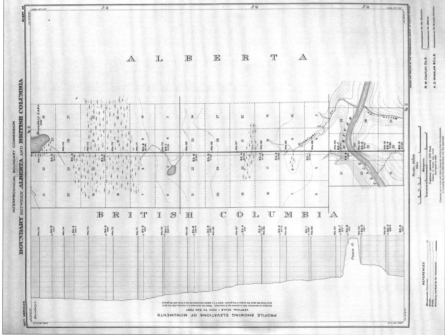

This is one of Cautley's map sheets for the 120th meridian. On the left side is a cross-section of the land showing the elevation. It shows the steepness of the Peace River valley in contrast to the rest of the terrain. On the right side is Cautley's map of the topography along the meridian line. His map includes the stations that he established, the sections of land the meridian line traversed, the geographical features, and contour lines for elevation. "Boundary between Alberta and British Columbia" map atlas, Sheet 49, Land Title and Survey Authority of British Columbia

The raft used to travel down the Peace River. e011205473, Library and Archives Canada

Cautley and his crew finished the field season at the south side of Boundary Lake, 21 kilometres north of the Peace River. Cautley had a large crew and a considerable amount of equipment. Instead of travelling to the rail line at Grande Prairie, he decided that it would be easier and quicker to build a raft and travel down the river. He described the craft the crew had constructed: "We made a raft of dry logs, thirty-two feet long and sixteen feet [10 metres by 5 metres] wide. On this, and on a small boat which we made out of whipsawed lumber, we all travelled about 150 miles [240 kilometres] down the majestic Peace River to railhead at Peace River Landing in six days." Meanwhile, the packers embarked with the horses on a journey of over 400 kilometres to Entrance, east of Jasper, where the animals would winter.

In 1919, Cautley and his crew completed surveying the first section of the 120th meridian south of Pouce Coupe. Then they worked north of this community, establishing the meridian through the main agricultural area of the Peace River region and crossing the Peace River in a deep canyon. Cautley's main objective was to survey the 120th meridian south to its intersection with the Great Divide, so he could not anticipate that he would return to Boundary Lake a few years later during the project.

Grande Prairie. j-01177,
Royal BC Museum and
Archives

SURVEYING THE 120TH MERIDIAN AND THE GREAT DIVIDE

1920

During this year, Wheeler's role in the boundary survey devolved to A.J. Campbell. While Wheeler remained boundary commissioner for BC and supervised the survey crew, A.J. Campbell assumed more responsibility for the actual survey work. Wheeler turned sixty in the spring, and he was finding mountain climbing more difficult. He was also starting a new venture.

Since working on the boundary survey around Mount Assiniboine during 1913 and 1916, Wheeler had envisioned developing a commercial walking tour that made a seven-day loop trip between Banff and Assiniboine. By the summer of 1920, Wheeler was ready to begin accepting clients. He received the necessary approvals from Banff National Park and established a base camp at Middle Springs in Banff. He also leased land from the BC government near Mount Assiniboine, initially for a seasonal camp. Wheeler urged the establishment of a provincial park to protect this area, and in 1922, Mount Assiniboine became a BC provincial park. To promote Wheeler's walking tours and tourism there, the Alpine Club held its 1920 camp at Mount Assiniboine.

Wheeler persuaded Ralph Rink to invest some money in the operation and act as the head packer. Jack Stelfox, who had worked on Wheeler's crew in 1913 and 1915, was hired as the cook. Frank Wallis, who had been assistant packer for the 1917 BC survey crew, was also hired. Wheeler gave Wallis some money and sent him on the train to Calgary to buy supplies. A couple of days later, Wallis's body was found by the train tracks near Canmore. An inquest was unable to come to any conclusion as to the cause of his death. Wheeler made no mention of this incident in his diary.

Involvement in the Alpine Club's summer camp, getting his commercial venture successfully started,

and overall supervision of BC's surveying and mapping work would keep Wheeler occupied during the 1920 field season. He had a replacement for the first part of the summer. Wheeler's son, Oliver, was going to India to survey for the British government, and he wanted to gain experience in phototopographic surveying. Oliver was also hoping to be selected as one of the men to attempt to reach the summit of Mount Everest, and he wanted to prepare by climbing peaks in the Rocky Mountains.

Spike Thomson was hired as second assistant and again kept a diary. Walter Nevler returned for a second season, along with Arthur Hughes, who became head packer. Mickey McNair was assistant packer, and Fred Weston, a New Zealander, was the cook. F.S. Archibald and V.G. Hunt were also members of the crew.

Thomson left Sidney on June 16. On the train between Vancouver and Banff, he met Thornton Andrews and his teenage son, Gerry. Over thirty years later, Gerry would become surveyor general of British Columbia, and Thomson would work for him. Arriving in Banff on the morning of June 18, Thomson spent two days working on Wheeler's 1919 surveying returns that needed to be completed.

The Boundary Commission report described a change in the route used at the beginning of the field season:

> It was intended to send the horses and some of the party and outfit to Jasper by way of the Bow, Saskatchewan and Athabaska Rivers route, but it was ascertained that, owing to the heavy snowfall of the previous winter and the slow rate at which the snow was melting, it would be impossible to cross Wilcox Pass which lies at an

Fortress Lake and Wood River. v771_pd_26_25_001, Whyte Museum of the Canadian Rockies

altitude of 7660 feet [2335 metres]. The balance of the party and outfit was to have gone by rail, via Calgary and Edmonton to Jasper, and there take over the food supplies shipped from Edmonton. In consequence of not being able to cross Wilcox Pass at that time, the entire party and outfit had to be sent round by rail at a greatly increased expense.

On June 20, Thomson travelled by train to Edmonton, and from there, he took the overnight train to Jasper. When he arrived the next morning, Wheeler and his son met him, and they "rode over to the Brewster Corral in sport model." The rest of the crew were already in Jasper, and they spent a couple of days preparing for the field season. Wheeler returned to Banff.

The crew left Jasper on June 24, heading back to Fortress Lake, where Wheeler had surveyed during the latter part of the 1919 field season. In 1920, the BC crew needed to finish surveying around Fortress Lake, carry the survey southward to link with stations surveyed during the first half of the 1919 field season and then continue surveying north along the Rocky Mountains.

The men arrived at Fortress Lake on June 27. Instead of using a raft to navigate around the lake, the crew cut out an old trail along the north shore. By July 1, they finished the trail to the west end of Fortress Lake, arriving at the previous summer's campsite along this part of the lake.

The men spent July 2 around camp. Thomson wrote: "I went out with the net & cyanide jar & caught some specimens for E.H. Blackmore, entomologist." Blackmore was the associate curator of entomology for the BC Provincial Museum of Natural History, and in his government report, he noted that Thomson collected a few specimens for him.

The next day, the crew moved camp about 3.2 kilometres down the Wood River, which flowed out of Fortress Lake, camping along the big flats where Alnus Creek joined the river. On July 4, Campbell, Oliver and Hunt surveyed from Fortress Lake West station while Thomson and Nevler located a trail up Alnus Creek, which flowed in from the north. The following day, the crew cut a trail down Wood River to the junction with Clemenceau Creek, which flowed in from the south.

On July 6, the men moved camp up Alnus Creek for about 1.5 kilometres. Then Campbell, Oliver and Nevler left for a fly camp. From this location, Thomson and Hunt surveyed a station the next day, while Campbell and his crew also occupied one before rejoining the crew. The men remained in camp the following day, then moved to a location along the Wood River near Clemenceau Creek. In 1919, the BC survey crew had surveyed the sizable Clemenceau Icefield from the south. Campbell now needed to establish stations from the opposite side.

Oliver departed for Jasper on July 10, accompanied by Hughes, who would bring back more supplies when he returned. Thomson, Nevler and Hunt surveyed from one station accessible from camp, while Campbell and McNair explored Clemenceau Creek to see if it was possible to take horses up this drainage. They decided that the slopes were too steep. The crew spent the next day cutting a trail up Clemenceau Creek, while July 12 was a rest day because the weather was uncertain.

The men started on a fly camp up Clemenceau Creek the following day. The Boundary Commission report described this trip.

> At the head of the valley is situated an exceptionally high snow-crowned peak of 12,004 feet [3659 metres] altitude above sea level. It is hung with glaciers and a wide icefield surrounds it on three sides. It was necessary to reach the southern borders of this icefield in order to cover the country that could not be reached the previous season from the head of the west branch of the Athabaska River. To do this the climbing party had to man-pack a camp up the valley, a most strenuous undertaking, and reach stations covering the watershed by crossing the wide icefield referred to.

That night, Thomson wrote, they "slept on a common brush bed with a log fire in front." The cook and assistant packer returned to the main camp after breakfast, while the rest of the crew travelled up the valley until early evening. On July 15, Campbell and Nevler surveyed from two stations on Clemenceau moraine. The next day, Campbell, Thomson and Hunt occupied two stations on the Clemenceau snowfield while Nevler stayed in camp. Fortunately, the weather remained clear, and they surveyed two more stations on July 17, completing their work on this section of the Rocky Mountains.

The men started returning to Wood River the next day. They were running low on food and supper was "mostly tea." By early next afternoon, the men were back at the main camp, where they spent the remainder of the day resting. July 20 was also spent in camp.

On July 21, Campbell, Thomson, Nevler and Hunt climbed Chisel Peak, located directly above the south shore of Fortress Lake. Wheeler had wanted to survey from this mountain in 1919, but had been unable to do so because of the smoke from forest fires. The elevation gain from Fortress Lake was over 1700 metres, and it took seven hours from camp to the peak. In the middle of their work at the station, "heavy thunderclouds passed overhead, forcing us to take shelter for

Clemenceau Icefield and Mount Tsar in left background. v771_pd_27_60, Whyte Museum of the Canadian Rockies

a while." The transit "began to sizzle & crack." Thomson observed that "Chisel seemed to be a storm centre—doubtless the influence of Mt. Clemenceau, the big white mountain to the SW [southwest]." In his additional notes, Thomson wrote that "during the heavy thunderstorm… one bolt hit the peak on the other side—the south—of the cairn. The flash was blinding, followed by a peal that rocked one's wisdom teeth. There was plenty of evidence that the peak had been struck repeatedly—there were fused rock here & there at the summit." After about an hour, the storm passed and the crew finished their surveying. They arrived back at camp after a 13½-hour day where they found that "Arthur had put up a Teslin stove for our tent for the first time." (This was a small portable stove that could be used for cooking and heat.)

On July 22, the crew moved to the mouth of Alnus Creek. The next day, they began cutting a trail up this drainage. The head of the valley lay along the Great Divide, and Campbell wanted to survey some

Mount Clemenceau. v771_pd_27_85, Whyte Museum of the Canadian Rockies

stations there. By July 28, the crew had cleared a trail and moved camp to the head of the valley. Thomson wrote about a difficulty that occurred during the trip.

> Shortly after our start I had a mishap with George, my saddle horse. We had crossed the creek to the north easterly side.... The trail lay along a narrow ledge between the water and a rock wall. Instead of dismounting I stayed in the saddle. About half way along the ledge, for some reason or other, George got off balance. I was thrown against the rock face of the canyon.... George disappeared into the turbulent waters. The last I saw of him were four hooves sticking out of the boiling pot as he was swept downstream. I thought, good bye to my personal belongings & my cyanide jar—not to say my lunch. Recovering somewhat from the unexpected, I decided to walk back down the trail. About a mile further down, to my surprise, here was George intact, being led by Art.... Apparently George had negotiated the swift water, mostly down on his back, until he got to the channel of the gravel flats below, when he regained his equilibrium and once more stood on his feet, none the worse for cold bath. Saddle–saddlebags–cyanide jar & lunch, tho slightly damp, had stayed with him.

The Boundary Commission report described the campsite and surrounding area.

Camp was pitched in a little grove of spruce near timberline that stood in the open grasslands almost at its head. It is a picturesque and beautiful spot. On the way up two fine glaciers send their icefalls directly into the valley and, in one case, the path passed directly by the foot of the ice which rises in a magnificent wall seventy to eighty feet [21 to 24 metres] in height.

Fortunately, the weather was good, and during the last three days of July, the crew was able to survey six stations in the area. Each day, Campbell took one crew and Thomson another. Thomson wrote in his diary that both of their stations on July 29 were along the divide and "from my set up, it was the first time I was able to look down into the Whirlpool Valley." At his station on July 31, Thomson was able to see Mount Fitzwilliam and Vista Peak to the northwest, where he had surveyed in 1917.

The crew returned to their campsite near the junction of Alnus Creek and Wood River on August 1, and the next day moved back to their June 27 campsite near the east end of Fortress Lake. One of the horses shook a rifle from its pack, and it took an hour to find it. Since it was wet and cloudy, the men remained in camp the next day. The crew left on the afternoon of August 4 to survey two stations located along the Great Divide north of their campsite. They camped that evening near the stations. Rising early the next morning, Campbell and Hunt surveyed one station, Thomson and Nevler the other. The work was completed by mid-afternoon and the men returned to the main camp in the evening.

On August 6, Campbell and the crew departed Fortress Lake, one of the few places along this part of the divide that provided fairly easy access to the BC side of the Rockies. The Continental Divide is convoluted through this area, and during their five weeks of surveying, the men covered the backbone of the Rocky Mountains to the south, north and northwest of the lake as it looped through this section.

Their next destination was Whirlpool Valley to the north. But first they needed to replenish their supplies for the remainder of the summer, so the crew headed towards Jasper. Two days of travel brought the crew

"A day off." 89.03.200, Jasper Yellowhead Museum & Archives

Camp in Alnus Pass. 89.03.223, Jasper Yellowhead Museum & Archives

to the junction of the Athabasca and Whirlpool rivers. The crew camped about 1.6 kilometres below the bridge at a site that the BC crew had used in previous years. Thomson called it "clean-up camp," since it was only a day from Jasper, and the men would clean up there before going into town. In his diary, he described an eventful evening.

> Cink—the Fire Warden, called in at camp about 7:35 pm to 7:45 pm. He returned about 9:30–10 pm as AJ & I were turning in. Called us to fight fire. He had located a fire about ¼ mile [400 metres] above the bridge over the Whirlpool. There was little doubt—it was a fresh fire & we were undoubtedly responsible for it. Somewhat reluctantly, we all turned out on parade.

It was getting on for midnight before we had collected up all available pots & pans, axes & what other paraphernalia we thought might be of use in firefighting. Having braced ourselves with hot coffee, the parade fell in behind Cink, the warden.... Fortunately the fire had not got away to a real start. It had not started to climb the trees, was confined to the moss & grass. We formed an endless chain to the river. Pots and pans, including the two large tea & coffee pots went into action. After several hours work, with the flames subdued, we crept around armed with what we had of water looking for sparks coming up out of the burn—quite easy to see on a pitch-dark night.... Finally

SURVEYING THE 120TH MERIDIAN AND THE GREAT DIVIDE

Raft constructed to cross the Whirlpool River. 89.03.214, Jasper Yellowhead Museum & Archives

figuring we had things sufficiently under control—returned to camp & made another round of coffee.

The next day, Hughes, Campbell and Thomson rode into Jasper while the rest of the crew stayed at camp to make sure the fire did not revive. In the late afternoon, the three men camped just outside Jasper by the junction of the Athabasca and Miette rivers, near where Thomson and the BC crew had stayed in 1917. After supper, the three men visited Colonel Rogers, the park superintendent, to report the incident. Campbell and Thomson spent August 9 shopping. In the evening, Hughes and Thomson visited Otto's pool room and also gathered more information about the Whirlpool and Athabasca area from the Otto brothers. The

next day, the three men returned to the camp on the Whirlpool River.

On August 11, the crew followed a well-defined trail up the west bank of the Whirlpool, a tributary of the Athabasca, for about 16 kilometres. Campbell wanted to survey some stations on the east side of the Whirlpool. Since they could not find a ford across the river, Thomson and Nevler spent two days constructing a raft, while other crew members cleared the trail going up the Whirlpool. August 14, the crew cut trail up the river, and the next day they surveyed a station near their camp.

The men remained in camp on August 16 because it was too smoky to survey and for the following two days because it was raining. On the evening of the 18th, the crew successfully launched the raft.

One of the photographs from the Mount Brown South station. v771_pd_31_261_001, Whyte Museum of the Canadian Rockies

The following day, the crew used the raft to cross the Whirlpool and surveyed a station above the east side of the river. While crossing the river, the raft had gotten stuck on a gravel bar. Campbell and Thomson disagreed over how to get it off safely, "the only time in all the years we worked together we clashed. How-ever, supper over, diplomatic relations improved & we turned in to our beds," Thomson wrote.

On August 20, the crew continued up the Whirl-pool to a location near the junction of the Middle Whirlpool. From their camp, they surveyed a station the next day, finding an old shell casing near the cairn.

Part of the panorama of Canoe Pass. v771_pd_31_279 and v771_pd_31_280, Whyte Museum of the Canadian Rockies

The crew travelled almost to Athabasca Pass on August 22. A short move the next day brought the crew to the Committee Punch Bowl, the famous tarn at Athabasca Pass. Thomson knew that this was a major fur brigade location in the nineteenth century, and he searched unsuccessfully for evidence. (In 1921, Cautley's crew would find artifacts.)

Inclement weather restricted surveying for the next five days. On August 24, Hughes left for Jasper, taking Hunt back to school and picking up Wheeler. The crew established a surveying target on a big rock near the centre of the Punch Bowl. Thomson and Nevler started up a mountain to survey a station, while Campbell and Archibald headed for another location. A storm arrived before the men could begin work, but they cached their instruments for a return trip. Rain kept the men in camp the next day. It was cloudy but not raining on August 26, so Campbell and McNair explored the area to the southwest, while Thomson and Archibald travelled up a narrow valley to the northwest, to a place named Canoe Pass, because the creek below them, on the west side, drained into the Canoe River, a tributary of the Columbia. It rained for two more days before

surveying resumed on August 29.

Meanwhile, Wheeler joined the crew for the first time. The Alpine Club camp was delayed for a week because the snow had melted late that summer. After the camp Wheeler intended to begin surveying, but when he arrived in Jasper, he received a telegram from Oliver that there were large forest fires in the Mount Assiniboine area. Wheeler returned to make sure that his clients, supplies and campsites were safe. Wheeler reached Jasper on August 25, and met Hughes when he arrived the next day. By August 29, the two men were at the camp in Athabasca Pass. In his diary, Wheeler wrote that the "boys put my tent up and made me comfortable."

On the same day, Campbell and Thomson returned to the stations they had been unable to survey on August 24. Though they completed their work, the two men were unable to read to all of the points because the high peaks remained in cloud. The next day, Wheeler, Campbell and Archibald surveyed from Mount Brown but did not complete their work because of clouds. Thomson and Nevler attempted to reach McGillivray Ridge, but by the time they approached the summit, the clouds were settling on the mountains. This station would be surveyed next year. On August 31, Campbell and Archibald completed their surveying. From the peak, Campbell saw Mount Robson on the northwest horizon. Meanwhile Wheeler, Thomson and Nevler surveyed a station on a ridge of Mount Brown.

On September 1, Wheeler surveyed from the target in the Punch Bowl while the rest of the crew cut a trail to Canoe Pass. The next day they moved camp to a small lake just west of the pass and were able to bring the horses along. Thomson built a survey target there. On September 3, Wheeler, Campbell and Archibald surveyed from the summit of Mallard Mountain, while Thomson and Nevler occupied three stations on Mallard Ridge, as well as surveying from the target near camp. In his later notes, Thomson added:

> At Sta. No. 1 disaster very nearly overtook me. I was set up on thin slab rock resembling slate. Going round the instrument, my foot caught a slab. It threw me against the instrument, knocking it over & sending me to the ground. On examination, I found that the tangent screw had been slightly damaged. With the aid of an elastic band, all went well.

Two more stations were surveyed the next day.

September 5 was Sunday. It was raining, with fresh snow on top of the mountains, so the men remained in camp. The next morning, Wheeler and Thomson read azimuths at the target, while Campbell and the rest of the crew began cutting trail. Wheeler intended to reach Fraser Pass, at the headwaters of the Fraser River, by travelling across the BC side of the Rocky Mountains. The Boundary Commission report commented: "The going was bad, and much trail cutting [was] necessary." Fortunately, the weather co-operated.

From their September 7 campsite, Wheeler, Thomson and Archibald occupied Fraser Pass SE, a station just northeast of the pass leading to the middle fork of the Whirlpool River. Wheeler wrote that it was a "glorious day for photography." It took four hours to reach the summit. Thomson wrote:

We ran into some tricky climbing about 500' [152 metres] from the top—shallow chimney, steep open face, with loose rock & rubble. We roped tho there was no belay at the top. We had tackled the southern slope, undoubtedly looked the best approach. On over our difficulty a quarter of a mile hike [400 metres] put us at our station.

When the surveying was finished, Thomson searched for a better route down but could not locate one.

> After a difficult bit of open fall I was last man—left high & dry on a ledge. AOW signalled to throw the rope down. I was on my own. Later, on the way home, he blasted me for my breach of mountaineering ethics for letting the rope go. Contrary blighter—I had already advised there were no belays on loose rock.

After this section of the mountain, there were steep parts but no real danger. The three men arrived at the new camp about 1.5 kilometres below Fraser Pass at 9:30 pm, covering the last section through the valley in the dark.

That night the weather changed dramatically and a large storm system started. Thomson celebrated his thirty-third birthday in camp, and Wheeler gave him a book for a present. By September 10, the precipitation turned to snow. On the twelfth, Thomson wrote in his diary: "All hands work on cook tent due to weight of snow." He also noted: "Snow settled in again at night, keep pushing snow away from tent walls." On September 13, the storm stopped, and the men brought supplies from their last cache. They took an inventory of their food and estimated that they were at least three days from the village of Lucerne near Yellowhead Pass. Campbell and Thomson climbed to the shoulder of a nearby mountain to look at the valley ahead.

In the evening, the storm resumed and continued for the next two days. On September 16, it turned to heavy rain. Thomson wrote that the ground was "becoming water-logged around the tents. Had to dig trenches." The next day it cleared sufficiently for Wheeler and Thomson to visit Whirlpool Pass. The Boundary Commission report stated: "It was ascertained that it would be advisable to make a monument survey of it, the easy approach from both sides of the watershed suggesting a suitable route over the Great Divide from the Athabaska River to the Columbia River." The two men also surveyed at the summit of Fraser Pass. Meanwhile, the rest of the crew brought the last supplies to the camp. Rain and sleet kept the crew in their tents on the next day.

The crew attempted to survey some stations on September 19. Wheeler, along with Campbell and Nevler, tried to reach a peak near Whirlpool Pass. Thomson and Archibald went to a station on a high ridge that overlooked this pass from the northwest. Snow on the ground made travel difficult. In his diary, Thomson described what transpired.

> We had no sooner reached the summit ridge when the storm set in—a blizzard from the SE [southeast]. We were knee-deep in snow. As the storm thickened, Archie & I became separated. Only by shouting from time to time were we able to keep track of our whereabouts. At the head of the gully I burrowed down in the snow—between big boulders—for about 6 feet or so [1.8 metres] until I was out of the wind—it was gale force. My next job to get Archie—continued shouting did the trick & I had him back to my dugout. Here we stayed for a couple of hours. However, with the possibility of being caught by darkness, we decided we would risk heading down the gully to the valley. This we did, more by luck, arriving in a meadow beside the creek about 4:30 PM. Lit a fire—warmed up. We set for camp 5:45 PM—arriving 6:30 PM.

Wheeler and his crew were also unsuccessful in reaching their station. There was more inclement weather the next day. Thomson wrote: "If the stormy weather continues, our food situation will become serious. We are already eating twice a day." They visited a nearby trapper's cabin and took some salt, since they had run out of this ingredient several days previously. (When they arrived at Jasper at the end of the season, the crew gave the trapper their remaining supplies, but this did not placate him for their taking supplies from his cabin.)

When snow continued on September 21, Wheeler decided to leave. The crew crossed Fraser Pass and began heading down the Fraser River drainage. By the time they reached camp about 3 kilometres below the pass, the precipitation became rain, and there was no snow on the ground. They found a dry, sheltered camp in the trees with feed for the horses.

The next day, the crew cut trail down the valley, while Wheeler went up to timberline east of camp to explore a valley that was supposed to lead to the North Fork of the Whirlpool River. There was snow on the ground on the morning of September 24, so the men remained in camp. When it cleared in the afternoon, Wheeler resumed exploring the area to the east.

It was snowing lightly on the morning of September 25. Wheeler realized that he would not be able to do any more surveying, so the crew headed down the Fraser River valley. They remained at their campsite a second night, because the horses scattered and not all of them were located until early afternoon. Meanwhile, Campbell took a group of men to clear the trail ahead. On the afternoon of September 28, the crew reached Lucerne, where Wheeler bought food, and the men then proceeded to Yellowhead Pass to camp for the night. The next day, Wheeler and his crew arrived at the campground by the Miette River bridge outside Jasper.

In summarizing the 1920 season, the Boundary Commission report stated: "It was a most exceptional season. The late melting of the snow in the spring and its early arrival in the fall, caused the failure of the division to close the gap and so finish the topographical work in the second section of the Boundary Survey, as had been intended." There were almost two weeks in August, between Fortress Lake and the Whirlpool River, where only one station was completed.

After September 3, Wheeler successfully surveyed only one station. He wrote: "By the time the weather

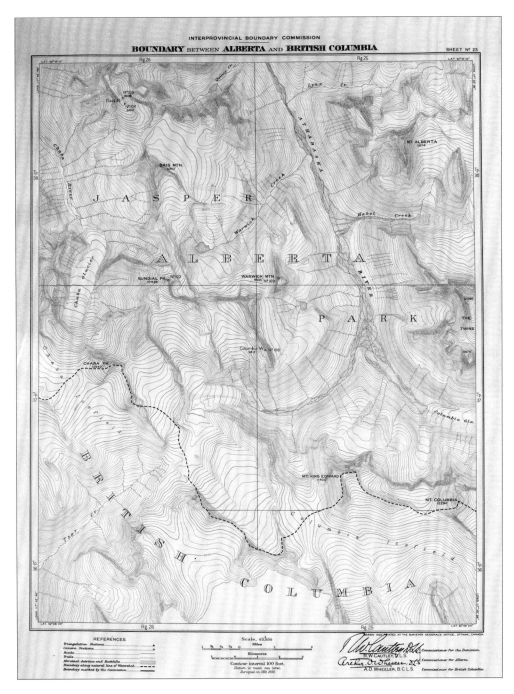

This map shows some of the area surveyed by the BC crew in 1919 and 1920, including Tsar Creek, where Wheeler made his mapping error. "Boundary between Alberta and British Columbia" map atlas, Sheet 23, Land Title and Survey Authority of British Columbia

cleared on the 28th of September, snow lay so deep on the mountains and on the area above timber line that it was impossible to reach high points for camera stations and the work had to be abandoned."

However, the September trip along the headwaters of the Whirlpool provided Wheeler with a better understanding of the geography of the area, and he would put that information to use in 1921. In addition, he had found a route that followed along the divide into the Fraser River watershed. Since the beginning of the Alberta-BC boundary survey, the Pacific side of the Rocky Mountains had drained into the Columbia River, but when he reached Fraser Pass, Wheeler entered a new watershed.

The 1920 season also produced the most serious error of the boundary survey. During the 1919 season, there had been few opportunities to survey along the BC side of the Great Divide for access was much easier from Alberta. The Boundary Commission report discussed this difficulty a couple of times. It observed that a station was occupied on the southern border of the Columbia Icefield "in order to supply some of the data for mapping, deficient through inability to reach the Bush River Valley," located in British Columbia. Farther north that same season, it became even more difficult to map the BC terrain. In describing the area around the headwaters of the Chaba River, the report stated:

> The continental watershed follows a circular ridge of numerous elevations enclosing the icefield of which the two main Chaba glaciers are the outlets.... What lies beyond it is, at the present time, a matter of speculation for it was not found possible to reach it from either of the Chaba branches. It is likely that information concerning the area lying beyond will be gathered next season by way of approach from the Wood River Valley.

In the conclusion of the 1919 chapter, the Boundary Commission reported on the mapping work.

The increasing difficulty of getting data on the British Columbia side of the Divide has made it impossible so far to map any part of it northwest of the Columbia Icefield until close to Fortress Lake. It is hoped to be able to gather data for such purpose during the season of 1920 by means of the snow pass at the head of the western source of the west branch of Athabaska River and of the valley of Wood River and its tributaries.

Circumstances changed for the British Columbia crew in 1920. The original goal had been to complete the phototopographic survey up to the 1917 survey of the Yellowhead Pass area, a full season's work. Since the men travelled to the Fortress Lake area (the farthest north they surveyed in 1919) via Jasper instead of Lake Louise, they did not have the time to go to the snow pass at the head of the Athabasca River. A.J. Campbell was in charge of the surveying in 1920. Wheeler had occasionally been criticized for mapping more than a few kilometres on either side of the boundary, so Campbell probably felt that surveying stations around the Clemenceau Icefield and from the Wood River valley would provide sufficient detail. It was possible to access both locations from Fortress Lake.

Describing the view from the icefield, the Boundary Commission report stated:

> Beyond the southeastern margin of the Clemenceau Icefield is seen the deep valley of a large stream which, judging by its general direction and source, may be the Sullivan River.… A tributary of this stream, Tsar Creek, heads from the snow-covered pass where the most western source of the Athabaska River has its rise.

Because the crew did not survey a station at the snow-covered pass, or anywhere in the vicinity, they did not gain a clear view of the geographical features on the BC side of the Great Divide in this area. A couple of stations were surveyed in the Wood River valley, but they provided only partial coverage of the BC terrain.

Nevertheless, the Boundary Commission reported in its mapping for 1920 that

> the season's surveys added the area southwest of the watershed to sheet No. 23, which could not be reached in 1919. It was now covered by the assistance of camera stations on the Clemenceau Icefield. The data obtained enabled the sheets to be mapped, producing a very nearly full sheet. The same stations enabled sheet No. 24, submitted in 1919, to be enlarged to practically a full sheet. In addition, new sheets Nos. 25, 26 and 27, filled out as fully as the data obtained permitted, are now submitted.

When Wheeler produced his maps back in Sidney, he unfortunately did not confine them to details that he could actually see on the photographs. Since he had not surveyed any of the stations, this was the only visual information that he had. Instead, to make his maps as full as possible, Wheeler interpolated contours and topographical features into terrain that the BC crew had not surveyed. He probably assumed that the western slopes of the divide would not have any glaciers. This was a serious error on his part and a blemish on his survey work. Wheeler's absence from the field was probably also a contributing factor.

Wheeler's error was first documented by A.J. Ostheimer, who climbed several peaks in the area in 1927. Ostheimer kept a journal that was the basis for a book titled *Every Other Day*. He wrote:

> The (A/BC) survey sheet showed that, in the valley labeled "Tsar Creek" there was timber and stream, but no glacier. On the contrary, in its waist is a glacier of considerable extent, measuring about ¾ mile [1.2 kilometres] across and flowing for some 3 miles [4.8 kilometres] from the névés at the valley head. For this ice river we have suggested the name of "Wales Glacier." As for the name of the stream flowing from Wales Glacier to Tsar Creek, we have suggested

the appropriate title, Wales Creek. Tsar Creek is misplaced on the survey map, for the name should belong to that creek with its deep and wild canyons that flow south and then SW from the glaciers of Mt. Tsar.

A footnote in the book states: "On the Boundary Survey Sheet #23, the Shackleton, Tsar and Wales Glaciers are absent; Wales Creek is called Tsar Creek; Tsar Creek is termed Sullivan River." In 1931, J.M. Thorington wrote in the *Canadian Alpine Journal*: "On the western side of the (Snow) pass, Wales Glacier, imperfectly mapped, descends to the depths of Tsar Creek." George Bell also wrote in the *Canadian Alpine Journal* that "the A/BC Interprovincial Survey Map is completely erroneous in the region 'Headwaters of Tsar Creek.'"

R.W. CAUTLEY

During this year, Cautley resumed surveying the 120th meridian south. The objective was to cut the meridian line, select the stations and supply them with the 900 kilograms of gravel needed for each monument. The surveying and construction of the monuments would be done in a future year. Cautley's assistant was D.M. Robertson, a young Dominion and Alberta land surveyor from Edmonton. Because of the large amount of work needed to cut the line, Cautley had a crew of eighteen men and a pack train of thirty horses.

Cautley had a difficult start to the field season because heavy snowfall and a late spring thaw produced extensive flooding in the Peace River region. In a letter to Deville from Grande Prairie on June 14, he described his problems.

All my outfit is here and I expected that my horses would have arrived on the 11th or 12th and that I should have been able to get away today.

On Saturday night, however, I received a wire from Chief Forester Badgley at Entrance, as follows: "your packers unable to force horses over Smoky. They are returning to Entrance. Will reach here about thirteenth." The Forestry Department have

telephone connection for a considerable distance along the trail, and that is how the packer was able to forward a message from Entrance before he arrived there.

Cautley also noted conditions in the area:

It has been raining for two weeks and, on the 11th instant we had four inches [10 centimetres] of wet snow. The main roads are under water in places, bridges and culverts have been washed out, and wagon travel is temporarily impossible....

The E.D. & B.C. Ry is in even worse condition than ever before, with washouts or wrecks occurring daily; when I came through last week we had to be transferred across a washout, and no freight was getting through, and when Mr. Robertson came through on the next train the engine left the tracks and the passengers were eventually brought on in a box car by the wrecking engine. Today's train from Edmonton and tomorrow's from here are both cancelled, and the agent cannot say when the line will be reopened.

Cautley was unable to make arrangements to ship his horses by rail to Grande Prairie, so he had to sell them in Edmonton and purchase a new bunch close to the Peace River. It took almost two weeks to get the entire outfit shipped to Grande Prairie.

In an interim report to Deville made in early August, Cautley described the slow pace of cutting the survey line during July: "I have not previously ever experienced 16 miles [26 kilometres] of such continuous heavy work." Timber appeared to be the only economic value for the land in the area.

Progress was slow: "There were no existing trails and it was found best to cut out the line itself wide enough and clean enough to make a trail for all the heavy packing that the building of concrete monuments entails."

Cutting line through dense brulé.
PA-018406, Library and Archives
Canada

Flooding along Bear Creek in Grande Prairie. PA-018403, Library and Archives Canada

The Boundary Commission report described the terrain:

> The country traversed by the 120th meridian from Monument 73-7 towards the mountains in the south is a high, gently undulating plateau with an irregular but constant rise to the south and west, so that each succeeding height of land crossed in a southerly direction is appreciably higher than the last.

During the 1920 season, about 70 kilometres of unusually heavily timbered line was cut for the 120th meridian, forty-four monument sites were selected, and thirty-three were supplied with gravel. The survey ended on September 28 at Monument 66-5 at the top of a 150-metre cut bank on the north side of the Wapiti River. Cautley and his crew reached Edmonton on October 6.

In the interim report Cautley made in October, he described the disappointing results and the fact that he had achieved only about half of his objective for cutting line during the 1920 field season. He attributed

SURVEYING THE 120TH MERIDIAN AND THE GREAT DIVIDE

Camp 13. Cautley wrote: "Space had to be cut in the brulé to set up the tents; there is no horse feed, and the horses cannot get off the trail for 3 miles [4.8 kilometres] either way from this camp on account of brulé." j-01175, Royal BC Museum and Archives

this to two causes: "shortness of working season and extraordinarily bad windfall." The delay in getting to his surveying in spring meant the loss of a couple of weeks of work, while the necessity of paying increased wages to obtain crew members meant that he had to return about a week earlier than usual. During the surveying, bad weather "caused a loss of working time this season considerably in excess of what is usually met with." Since only one source of gravel was located, there was considerable work transporting the material to the monument sites. The Boundary Commission report noted that the entire country traversed by the meridian was "heavy windfallen brulé to a height from three to seven feet [1 to 2.1 metres], interwoven with dense, small second-growth jackpine." (Brulé was a forested area destroyed by fire. The windfall from the burnt trees often made it difficult to travel through the brulé and to cut the survey line.)

Signboard at Yellowhead Pass. 89.03.62, Jasper Yellowhead Museum & Archives

SURVEYING THE 120TH MERIDIAN AND THE GREAT DIVIDE

1921

The Boundary Commission planned to complete the second section of the survey, from Kicking Horse Pass to Yellowhead Pass, in 1921. In addition to the phototopographic surveying done by the BC crew, Cautley and the Alberta crew needed to monument some of the passes in this section. After 1917, when the Alberta crew had delineated Howse and Yellowhead passes, no work had been done on this section. Except for Yellowhead, the passes along this part of the Great Divide generally did not have economic or geographical significance. However, the commission believed that more of the passes should be surveyed and monumented, so Cautley spent the 1921 season doing this work.

Following discussions with all of the main people involved in the boundary survey, four passes were selected. Fortress Pass was chosen because of its unusual geography. Athabasca Pass, which was actually at the headwaters of the Whirlpool River, a tributary of the Athabasca, was selected because of its historical significance. Whirlpool, at the head of the Middle Fork of that river, was near Athabasca Pass and could be easily surveyed while Cautley was in the area. The final pass chosen was Tonquin. On the BC side, it was on the Fraser River drainage, while on the Alberta side it was at the western end of the scenic Amethyst Lakes valley, a popular recreation area in Jasper National Park.

As had been the case in previous surveys of the passes, Wheeler would need to spend time with Cautley making sure that he agreed with the proposed survey line through each of the four passes. He would also have to spend time with Campbell and the BC crew, for it was important that the survey proceed smoothly, and that the stations covered all of the remaining areas along the second section of the Great Divide. In addition, Wheeler planned to continue his walking tours to Assiniboine, and he needed to provide some supervision there.

Spike Thomson did not return, so Wheeler had to hire a qualified assistant to work with Campbell. W.J. Moffatt, BCLS #188, joined Wheeler's crew. Wheeler took Jack Stelfox from the walking tours to assist him in his travels between the crews during the summer. Stelfox could handle horses, cook, and help Wheeler with surveying.

The logistics for the 1921 field season were complex. Cautley, Wheeler and Campbell had to co-ordinate their surveying so that each person worked as efficiently as possible. The pack horses had to move men and supplies in a timely manner. Communication would be limited in the remote areas where each group was working, so Wheeler and the two crews would have to stay on schedule as much as possible and hope that no unforeseen circumstances would arise. Both Cautley and Wheeler kept diaries, and they provide details about the surveying that each person did.

Since Cautley had the most work to accomplish, the Alberta crew needed to get into the field first. The men planned to start at Fortress Pass, the most southerly of the four. Cautley had to get a preliminary line surveyed by the time Wheeler arrived, and geographically, it was the most detailed of the passes to be surveyed in 1921.

Cautley had an immediate problem. In 1920, high water had prevented the packers from bringing his horses from the Jasper area to the Peace River country. At the end of the 1920 season, Cautley anticipated returning to the Peace River region in 1921, so he left his horses at Grande Prairie for the winter instead of taking them back to Jasper. Now, at the beginning of the season in mid-May, there was too much snow in the mountains, preventing the packers from moving the horses south. Cautley had to go to Grande Prairie

Fortress Pass

Fortress Mt.

28   29   30 31

Fortress Lake
Draining To Columbia river + Pacific

Summit
f
Pass +
Boundary
between
Alberta + B.C.

Chaba River
draining to Arctic

A panorama of the Fortress Pass area that shows the Great Divide along the narrow strip of land between the Chaba River and Fortress Lake. PR1982.0230.0002, Provincial Archives of Alberta

and arrange for the horses to be shipped to Jasper.

On May 16, Cautley left Edmonton for Grande Prairie, accompanied by the head packer, J.F. Condor. The ED & BC railway lived up to its reputation. The next day Cautley wrote that the train was stalled at McLennan for several hours before proceeding to Spirit River at 4:30 pm. Cautley and Condor arrived in Grande Prairie at 8:30 the next morning. The train derailed five times in the 80 kilometres between the two towns.

At Grande Prairie, Cautley went out to the ranch where the horses had spent the winter, but the rancher was away looking for lost horses, so he left word to bring them into town for shipping the next day. When the horses arrived on May 19, he found two mares were in foal, so he traded them for two saw horses. Another horse was in poor condition and had to be shot. Altogether, twenty-one horses were put in Condor's charge

for shipping, while Cautley returned to Edmonton. In his diary entry for May 20, Cautley wrote that the train derailed once while he was awake, and he thought it also happened a few times while he was asleep. He arrived in Edmonton the next day.

Cautley completed arrangements for the field season over the next ten days. He finished hiring his crew and ordering supplies, which he shipped to Jasper. On May 27, the horse car arrived from Grande Prairie, and three days later, the animals were sent to Jasper. Part of the crew left for Jasper on May 31, and Cautley, along with the rest of the men, departed the following day. The crew spent two days at Jasper getting the outfit and horses ready, along with storing items needed later in the season at the Brewster warehouse. The crew spent the evening of June 3 at the Miette bridge camp just outside Jasper.

The initial two days of travel went smoothly, but

on June 5, Cautley met a park warden who told him that the water level of the Sunwapta River was high and still rising, and he advised Cautley against crossing the river there. The next day, Cautley tested the ford (which was not far above the confluence with the Athabasca) and concurred with this warning, so on June 7, the crew moved up to Sunwapta Falls. Cautley wrote:

> In afternoon commenced bridge of 30 foot [9 metres] span between supports in the narrowest part of extraordinarily wild canyon with sheer rock walls 100 feet [30 metres] high. Water still very high. The bridge has to be built in very awkward place, but the men are surefooted and plucky.

They completed the bridge and installed a handrail the next morning, then crossed the bridge and cut 3 kilometres of new trail towards the main trail up the Athabasca River. The next morning another 1.5 kilometres of new trail brought them back to the main trail, and by June 10, Cautley and his crew had crossed the Chaba River and established a camp about 400 metres from Fortress Pass.

Wheeler came to Banff in May, and spent most of his early preparations on the walking tours. He hired Stelfox and F.S. Archibald to assist him with a variety of activities for both the Alpine Club and his business venture. In early June, Art Hughes returned for another season as head packer for the survey crew, and Campbell arrived. On June 9, the pack train with Hughes, the assistant packer, Jimmy Lamb, and Campbell left Banff for Jasper. With Ralph Rink, Wheeler then went on a four-day trip to visit some of the facilities along the route of the walking trails. Then he finished the report and maps for the 1920 surveying and made preparations for the upcoming season. Wheeler left Banff on June 23 with the rest of his crew, spent June 24 in Edmonton buying supplies and arrived in Jasper on June 25, where Campbell and the packers met him, having arrived three days earlier. The Otto brothers hauled Wheeler's gear to their camp, where he put up his tent. Wheeler noted that Cautley's packers, who had come

Bridge constructed over Sunwapta Gorge. e011205446, Library and Archives Canada

into Jasper to pick up cement for the monuments at Fortress Pass, were camped at the Miette River bridge.

Wheeler was anxious to get started, and on June 26, he awoke at 4:00 am and called the packers to get the horses. However, the horses were not co-operating, and six hours later, the packers had located only nine of them, so Wheeler and Stelfox left, taking two saddle horses and two pack horses. They travelled about 25 kilometres, stopping to camp around supper. Stelfox wrote an account of an adventure that occurred the next morning.

> Finishing breakfast, I went to round up our horses, leaving Mr. Wheeler to have a smoke before breaking camp. Returning with the horses, I found Mr. Wheeler asleep, and a black bear with its head in the porridge pot. He was too busy to notice me as I crept up behind him and clouted him on the rear with a club, much to his disgust. I later found out that the warden, McGuire, had two pet bears that stayed around his cabin nearby.

The two men crossed the Sunwapta on Cautley's bridge, and travelled until about 7:00 pm. On June 28, Wheeler and Stelfox got an early start. En route to Fortress Lake, they met Bruce Otto and a timber cruiser.

While travelling through some timber along the riverbank, Stelfox had difficulty with his horse, in an incident he described.

> My saddle horse got scared, jumped between two trees only about 3 feet [0.9 metres] apart. Each of these caught my legs and lifted me out of the saddle. I took a nasty spill, which made me quite mad. My saddle horse landed in the river, and after being carried quite a distance down the river (the current was quite strong here and lots of white water) climbed out on the far side and proceeded to try to buck off the saddle and a sack with cowbells used on the horses when grazing around camp.

Stelfox had to get his horse back, and he persuaded Wheeler to let him use his horse. Stelfox swam the river with Wheeler's horse, roped his own horse and tied him to the saddle, and swam back across the river.

> By this time Mr. Wheeler was ready for one of his cigars. He carried a supply in a poncho on the saddle and his lunch in the other saddle poncho. Naturally, during the swim both ponchos had filled with water, and both cigars and lunch were too wet to be any use. Then I got a sample of an Irishman's tongue.

Stelfox insisted on stopping to have a hot meal for lunch, much to Wheeler's displeasure.

> I was not long in preparing a hot lunch, which was eaten in silence, but I think it made both of us feel better, for Mr. Wheeler was soon digging into his dunnage bag, and came over to me with a handful of cigars and some oranges and said, "you and I are going to travel a long way together."

Stelfox then commented: "Since that trip, I have made many trips with that fine old mountaineer." Wheeler arrived at the Ottos' Fortress Lake camp that evening and visited Cautley. By this time, Cautley had almost completed locating the boundary line through Fortress Pass.

In 1922, Cautley wrote an article for the *Canadian Alpine Journal* titled "Characteristics of Passes in the Canadian Rockies." It included a description of Fortress Pass.

> Considered technically, from the point of view of watershed determination, Fortress Pass… is the most extraordinary of all the main divide passes so far dealt with. It lies between the Chaba River, which at this point is a large stream flowing in many channels through one of those wide gravel beds so characteristic of glacial streams, and Fortress Lake, which is a fine lake

six and a half miles [10 kilometres] long and hemmed in by high mountains. The summit of the pass consists of a timbered flat about a quarter of a mile [400 metres] square between confining mountains and the river and the lake referred to. The flat has a constant and uniform grade from the river towards the lake equal to the grade of the river itself, the water level of the lake being about nine feet [2.7 metres] below that of the river. As no point of the flat along the riverbank is more than 3½ feet [1.1 metres] above the water level of the river, while the flat along the lake shore is nearly nine feet below it, the situation is truly remarkable and there can be no doubt seepage from the river to the lake actually takes place at all times of high water in the river.

After the Alberta crew's arrival on June 10, rain on June 11 limited work, and the next day was Sunday. In his June 13 diary entry, Cautley wrote that he "ran various preliminary lines & took levels over them.... The flat between the Chaba and Fortress Lake is an extraordinary case of watershed." The following day he described his surveying.

> Did excellent day's work & practically completed cross-sections of flat, which means a whole lot of work as it is heavily timbered throughout. Plotted cross-section and now know where the watershed line is. The highest point is only 3'6" [1.1 metres] higher than the water of the Chaba which is opposite to it.

On June 15, Cautley and his crew spent the day locating the divide along the south side of the pass. "Watershed is flat and ill-defined and covered with heavy timber up to 36 inches [0.9 metres] diameter." The next day, while Robertson and the crew ran the preliminary line along the south side of the pass, Cautley and the monument-builder, Mark Platz, climbed about 775 metres up the mountain, looking for the location of the

This map shows the Alberta-BC boundary through the Fortress Pass area along with the Chaba and Clemenceau icefields. "Boundary between Alberta and British Columbia" map atlas, Sheet 24, Land Title and Survey Authority of British Columbia

highest station on the south side of Fortress Pass. On June 17, Cautley and the crew started a preliminary line on the north side. He noted "watershed somewhat involved and tremendous green timber." Over the next three days, the Alberta crew cleared the line between

the stations on the south side and finished the preliminary survey. In his diary entry for June 20, Cautley wrote: "Bear tore cover of my transit at 3-O [the letter designation for Fortress Pass is O], but no damage to the transit."

On June 21 and 22, Cautley began surveying the trial final line between the stations closest to the centre of the pass. The next day, he continued the preliminary survey on the north side of the pass, up towards Fortress Mountain: "Heavy timber & very steep hill side broken by small cliffs made progress slow, and I did not get all the information I hoped for." During that time, the packers left for Jasper to bring back the bags of cement needed for the monuments.

Over the next four days, Cautley completed the final trial line on the south side and cleared a trail to get the horses and cement as far up the pass as possible. Cautley found that a bear had torn a signal to shreds. The crew also began the final trial line on the north side. They had partially completed this work when Wheeler arrived on June 28.

Wheeler and Cautley spent the last two days of the month going over all the lines and proposed monuments on both sides of the pass, while Robertson continued working with the crew and Platz began preparing the monuments. Wheeler set a bolt and built a cairn for the highest station on the north side of Fortress Pass. The packers arrived with the bags of cement.

As was the case with all the previous passes that Cautley had surveyed (except for Phillipps in Crowsnest Pass), Wheeler concurred with Cautley's work.

On the afternoon of July 1, Wheeler left, going back to check on his walking tours venture. While returning to Jasper, he stopped to visit Campbell and his crew, which had departed for the Athabasca valley a day after Wheeler and Stelfox. In his diary entry for July 2, Wheeler wrote: "Made Campbell fly camp at 1 PM. Had lunch and tea there. Campbell and crowd came down from mt [mountain] when I was at his camp. Gave him instruction. Went on at 2:30 PM."

Top: Hooker Icefield and Mount Hooker, back right. v771_pd_33_33, Whyte Museum of the Canadian Rockies

Bottom: Mount Hooker, back right, and area. v771_pd_33_41, Whyte Museum of the Canadian Rockies

At Fortress Pass, Cautley's packers brought up gravel and the construction of the monuments started, while Robertson and the rest of the crew finished work at the top of the pass on both sides. Now the final survey of the pass could commence.

Rainy weather during the first week of July slowed progress, but work continued. The distances between the stations had to be chained and elevations read. Construction of each monument was a slow, multi-step process. At night, Cautley made observations on stars to connect with the surveying. By July 9, Platz had finished constructing the seven monuments through the pass, and then he needed to paint them. Cautley and one man climbed to a triangulation station on a bluff near the top of the north side, where they read azimuths and elevations.

July 10 was a Sunday. The following day Cautley surveyed and took photographs of the monuments. Robertson and one man did more chaining, while the rest of the crew finished cutting the final line on the north side of the pass. On July 12, the last details were completed on the south side of the pass. The next day Cautley and Robertson "worked out and calculated all observations to date." Robertson did some final surveying at the monuments on the south side, while the crew finished cutting the survey line at the top of the pass on the same side. In the evening, Cautley recorded observations of the stars. Cautley had now completed the boundary survey of Fortress Pass.

On July 14, Cautley's crew started for Athabasca Pass, arriving five days later. The trail up the Whirlpool River was in bad condition, making travel slow. At the pass, they found A.J. Campbell and the BC survey crew. From June 27 to July 9, Campbell's crew had surveyed four stations in the Athabasca and Sunwapta valleys. They then proceeded to the Whirlpool River and travelled up to its junction with the Middle Whirlpool. From that location, on July 15 they took a fly camp to Athabasca Pass, where they surveyed two stations around Mount Hooker that had not been completed in 1920 because of inclement weather. Cautley met Campbell and his crew on the evening of July 19 as they were completing their work, and they left the next afternoon.

To survey and monument Whirlpool Pass, Cautley needed a suitable trail for taking the horses up the

Middle Whirlpool River and Mount Edith Cavell in centre left background. v771_pd_34_96, Whyte Museum of the Canadian Rockies

Middle Whirlpool River, because the animals had to pack the material needed for the monuments, along with equipment and food. Campbell also had to take his crew up to the pass. At the beginning of the season, arrangements had been made for the BC crew to cut the trail to Whirlpool Pass, which they did from July 21 to 30. The Boundary Commission report described this arduous work.

> For the first three miles [4.8 kilometres], after leaving the Whirlpool River, the Middle Whirlpool flows through a deep cañon with very steep sides, so that the trail rises about 1500 feet [450 metres] very rapidly. After the first three miles the grade is easier, but rock falls and high earth cut banks made trail building difficult. Between four and five miles [6.4 to 8 kilometres] from the summit there are extensive beaver meadows with excellent horse feed, but there is no feed closer to the summit except above timberline.

The Jasper Yellowhead Museum has a couple of the musket balls found in 1921 on display. Photo by Jay Sherwood

After completing the trail, Campbell spent a week surveying five stations around the pass.

On July 20, Cautley started surveying Athabasca Pass. It was easy to determine the location of the Great Divide through this location. In his article about the passes of the Canadian Rockies, Cautley wrote that Athabasca "has a lake named the 'Punch Bowl' on its summit, from which water flows both towards the Arctic and Pacific Oceans." The Boundary Commission report provided more detail about the determination of the boundary in the pass.

> Athabaska Pass is the best example, so far dealt with by your Commissioners, of a mountain pass of which the actual summit is a small lake, with a visible flow both towards the Arctic and Pacific oceans. This small lake which… is known as Committee Punch bowl, is nine chains long in the direction of the pass and six chains broad [180 metres by 120 metres].…
>
> Before definitely deciding to treat Committee Punch Bowl as a summit lake, and therefore the key of watershed determination, your Commissioners exam-

> ined it together, very carefully, and found: first, a small but perfectly obvious stream flowing from its northerly end into Alberta; secondly, a small flow into British Columbia from its southerly end.… Your Commissioners particularly desire to put themselves clearly on record in regard to the above facts, because it seems probable that the conditions found by them in July, 1921, may easily be found quite otherwise by subsequent visitors to the pass.

The report noted that both streams could easily become dammed or affected by rockfalls from a steep mountain that rises above the lake. It also observed that "Athabaska Pass is essentially a bare rock pass" and that "all of the six monuments in Athasbaska Pass are built upon solid rock."

In his *Canadian Alpine Journal* article, Cautley also described the historical significance of the pass, which was known to the Indigenous inhabitants of the area and had been used by David Thompson. This history became a reality during the Alberta crew's work in Athabasca Pass: while searching for gravel for the monuments he was building, "Mark [Platz] discovered [a] cache of 114 old musket bullets," Cautley wrote in his diary. Cautley later found a reference in Thompson's account of crossing Athabasca Pass for January 13, 1811: "Sent the men to collect and bring forward the Goods left by the Way; which they brought except five pounds of Ball, which being in a leather bag was carried away by a wolverine."

On July 21, Cautley explored the pass on the west side, while Robertson and the crew located the watershed and started running trial lines near the pass. Platz began finding and washing gravel for the monuments. The packers left for Jasper with nineteen horses to pick up supplies. And with impeccable timing, Wheeler arrived in the evening.

Stelfox had met Wheeler when he arrived in Jasper on the morning of July 19.

> Jack had horses there & he took my things to the old camp where I changed and sorted things out. Jack then took my town

things back to Otto's and about 12:20 we started for Whirlpool Bridge. Got there about 6 PM. Camped—slept under trees & I suffered from mosquitos.

With an early start the next morning, Wheeler and Stelfox reached Campbell's main camp at the junction of the Middle Whirlpool about the same time that Campbell arrived from Athabasca Pass. In the evening the two men discussed the surveying Campbell had done. On July 21, Wheeler and Stelfox left for Athabasca Pass, accompanied by Jimmy Lamb, the assistant packer. Along the trail they met Cautley's packers, who were heading for Jasper. Arriving at the pass in early afternoon, Wheeler set up camp and spent the evening visiting with Cautley. He agreed to send eight of his pack horses to Jasper to bring supplies for the Alberta crew.

The next morning Lamb returned to Campbell's camp with a letter instructing Hughes to go to Jasper for Cautley's supplies. Wheeler and Cautley went out to inspect the boundary line through Athabasca Pass, while Robertson and the crew completed the preliminary survey. Then Wheeler took Stelfox and Red Burt from Cautley's crew and "went up to ridge south of Mount Brown and put bolt in rock block on skyline at point and after for boundary." The next day the three men "climbed up face of Punch Bowl Ridge. Set bolt & built cairn." Cautley and Wheeler had supper together that evening. On July 24, Wheeler left for Jasper, stopping for lunch at Campbell's camp and remaining for over two hours. About 8:00 pm, he and Stelfox stopped to camp, but the mosquitos kept him awake for several hours. The next afternoon he arrived at the Miette bridge at 5:15 pm, where he found Hughes and had a meal. Two hours later, Wheeler was on the train to Edmonton.

Wheeler departed on Sunday, a rest day for the Alberta crew. In his diary, Cautley wrote: "Fine day. Whole party had swim in Punch Bowl." Work resumed on July 25: "Climbed about 2600 ft [800 metres] to 10-P. Read very good azimuths & elevations. Built good cairn.... One piece of nasty rock climbing—almost 75 feet [23

The Committee Punch Bowl in Athabasca Pass. e011205447, Library and Archives Canada

scheduled to return, it rained hard all day. On July 31, late in the evening, the packers finally arrived.

The first week of August was spent finishing all the details of surveying Athabasca Pass. On August 1, Cautley marked the brass plates, the men packed 270 kilograms of monument material up 180 metres of bluffs beyond the horse trail to Monument 8-P, and Platz began building the monuments. While the work continued the next day, Cautley constructed some signals to use for his surveying. Rainy weather limited work for the next two days. In his diary, Cautley described a busy day on August 5: "Mr. R. doing precise chaining. Self read azimuths & observation at 8-P, 6-P and Δ Sta C [triangulation station]; also photographed. Mark built 4-P and 1-P and put on top of 2-P. Observed at night. Men moving instruments & monument forms, etc. Also ran out line from 1-P to 3-P." On the next day, which was similarly busy, Cautley and the crew completed work in Athabasca Pass.

After Sunday in camp, Cautley's crew left on August 8, headed for the camp near the junction with the Middle Whirlpool River. The next day they proceeded up this river valley and

> arrived in Whirlpool Pass in thunderstorm, very late and horses really played out. It is about 9 miles [14 kilometres] from where Mr. Wheeler's trail leaves the main river up to the summit & his outfit have done a great deal of work on it, otherwise it wouldn't be possible at all. As it is, it is very steep in places and a difficult stony trail. No feed within 4 miles [6.4 kilometres] of summit.

Campbell's BC crew had departed the pass earlier in the day. They were working north towards the Fraser River watershed along the BC side of the divide, following the route used in 1920.

Cautley and his crew spent the first two days at Whirlpool Pass locating the divide and exploring the area. The packers went back to the camp on the Whirlpool to bring up the supplies that Wheeler's packers had cached there. Then they went to Jasper to procure additional supplies. Platz searched for gravel. On Au-

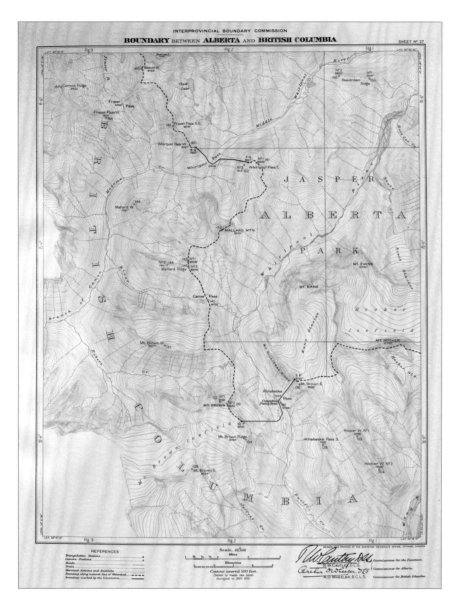

This map shows the location of Athabasca and Whirlpool passes along with some of the locations surveyed in 1920 and 1921. "Boundary between Alberta and British Columbia" map atlas, Sheet 27, Land Title and Survey Authority of British Columbia

metres]—otherwise an easy climb. Observed in evening. I let men climb to summit of Mount Brown while I was busy."

On July 26, the crew cut out the final lines, and Platz got the monument material ready. After finishing a few details the next day there was "nothing more for men to do until monument material arrives on Thursday—I hope." The next day, when the packers were

　SURVEYING THE 120TH MERIDIAN AND THE GREAT DIVIDE

Packing monument material to a station in Athabasca Pass. e011205448, Library and Archives Canada

gust 12 and 13, the crew ran a preliminary survey on both sides of the pass. On Sunday, August 14, several men went down the Middle Whirlpool to fish. They returned with thirteen trout, the largest weighing 1.4 kilograms. Work resumed the next day. Cautley described the activities on August 16.

> Fine day but smoke very bad. I can smell it and cannot see nearby peaks. Self and Mark climbed 2500 feet [762 metres] going up to summit of easterly ridge and setting signal as possible position of 7-Q, then travelled south over glacier and set another signal at C near foot of Mallard [Peak]. Mr. R. and men cutting final line to 4-Q. Observed in A.M. and evening.

Wheeler arrived on the afternoon of August 18. The last three weeks had been eventful for him. When he left Jasper on July 25, he was on his way to the Alpine Club's summer camp at O'Hara Meadows. But on the train between Edmonton and Calgary, he read the newspaper and learned about a fatal climbing accident on Mount Eon near Mount Assiniboine. Upon arriving

Fish caught on Middle Whirlpool River. 89.03.76, Jasper Yellowhead Museum & Archives

in Banff on the evening of July 26, the park superintendent met Wheeler "and turned over the matter of accident to me." He stayed the night at his walking tours camp at Middle Springs and was "up for quite a while talking things over with Clare," he wrote in his diary.

Dr. Winthrop Stone was president of Purdue University in the United States. He and his wife, Margaret, were avid mountaineers who had previously climbed in western Canada. They attended the 1920 Alpine Club camp at Mount Assiniboine and returned in 1921, staying at Wheeler's walking tours camp before starting on an unguided ascent of Mount Eon. The couple were almost at the summit on July 17 when Dr. Stone unroped to check the route. He fell and died. Margaret attempted to descend by herself, but got stuck on a ledge and was trapped there for seven days before being rescued on July 24. When Wheeler arrived in Banff, neither Mrs. Stone nor Dr. Stone's body had yet been brought out to Banff. During the next three weeks Wheeler co-ordinated the rescue and recovery effort

and also spent time at the Alpine Club summer camp. He took unpaid leave for the additional time that he was away from the boundary survey.

Wheeler arrived in Jasper on August 16 where Stelfox met him. The two men reached Whirlpool Pass two days later during a period of inclement weather. Wheeler "had supper with Cautley. Spent evening in his tent chatting." The next day he and Cautley inspected the stations on the east side of the pass and determined the location of station 5-Q. From there Wheeler went with Red Burt to Station 7-Q, the highest one on the east side of the pass, and set a bolt in the rock. He also visited Campbell's phototopographic stations in the pass. The following afternoon, after the rain stopped, Wheeler went with Cautley to examine the stations on the west side. Sunday, August 21, was a rest day, and it rained. Wet weather kept the men in camp the next day, and the packers returned from Jasper. Wheeler noted that there was new snow on the mountains. On August 23, "it looked like clearing so we started up Whirlpool West

Fraser Pass. 89.03.160, Jasper Yellowhead Museum & Archives

with Mr. Wheeler and men and instruments," Cautley wrote. "It began raining again as soon as we left camp and after remaining out 4½ hours on the chance of it clearing, I cached the instruments and came in." Wheeler took Stelfox and placed a bolt on station 6-Q at the top of the west side of the pass, but he was unable to survey at the station. After lunch the next day, Wheeler and Stelfox departed, heading to Campbell's camp somewhere near Fraser Pass.

On August 25, Robertson and Charlesworth surveyed at 7-Q, while Cautley and two men returned to 6-Q:

> Just after reaching the summit, it began to snow so that we couldn't see 50 feet [15 metres] for two hours, but cleared up partly so that I was able to see enough points of my survey, although AOW's cairn remained in clouds. Built cairn at 6-Q. Mr. R. was driven down by snow without being

able to do what was wanted.

Cautley observed that the next day was the first one with fine weather in twelve days. He spent the day surveying while Robertson completed the work on 7-Q and Platz built monuments. Excellent weather continued from August 27 to 29, enabling Cautley's crew to complete the survey of Whirlpool Pass during that time.

In describing the pass, the Boundary Commission report stated: "Whirlpool Pass summit is a narrow rock-strewn pass over the watershed, which the centuries have succeeded in clothing with scrub timber." The report also noted: "As far as is known by the Boundary Survey, Whirlpool Pass has no history." The commission did not believe that Whirlpool would be a significant pass: "It is never likely to have much importance as a route of travel, because the approach from the Alberta side is such that it would be difficult to construct a good road." In addition, "the valley floor, both at the summit and at many other places, is composed

Glaciers at head of Geikie Creek. v771_pd_37_240, Whyte Museum of the Canadian Rockies

of loose rockfall, thinly covered with moss and earth, over which it would be very expensive to make even a fairly good trail." Five bolts and two monuments were constructed to mark the boundary in the pass. It was not a large pass to survey, but Cautley's crew was there for three weeks because of inclement weather and Wheeler's delay in reaching the location.

The crew departed on August 30, heading for Tonquin, their final pass. Cautley described the day: "Packed out to Whirlpool River flats in pouring rain all the way, rain beginning just as the last horse was packed. Very bad trail." There were showers during the last day of the month as the men continued their journey. On September 1, Cautley wrote in his diary: "Poured rain all night and did not move today as the whole outfit is wet, ropes, horse blankets etc. Showered all day." The next day the crew reached the mouth of Whistler Creek near Jasper, where the trail to the Amethyst Lakes and Tonquin Pass started. Since Tonquin Pass was in subalpine terrain, Cautley paid off two axemen at this time. Robertson and the crew started up Whistler Creek, while the packers went into Jasper for supplies. Cautley also went to Jasper "to arrange for wagon & supplies & get them delivered in time for Monday packing."

The trip to Tonquin Pass started on September 5.

Packers took full load of oats and cement about 3 miles [4.8 kilometres] towards Marmot Pass and 2200 feet [670 metres]

SURVEYING THE 120TH MERIDIAN AND THE GREAT DIVIDE

above camp & made cache. Men made log cache at camp. As it will be possible to start building monuments just as soon as we reach Tonquin Pass, owing to open nature of pass, I am double tripping in and taking cement, etc. with me.

After three days, Cautley stopped for a rest day on September 8: "Snowed last night and this morning. The horses are played out with three days climbing and felt obliged to give them one day's rest." The next day the crew reached Tonquin Pass. Along the way, at Amethyst Lakes, they met Campbell's party heading out.

In the month since leaving Whirlpool Pass, Campbell's survey crew had spent most of their time along the headwaters of the Fraser River. The Boundary Commission report stated: "From August 12 to September 2nd the phototopographic party occupied camera stations on both sides of the Fraser Valley during such time as the weather permitted. Bush fire smoke and, later on, rain and clouds interfered greatly with this part of the work."

Then Campbell went to Geikie Creek, a tributary of the Fraser. From September 5 to 8, he surveyed at five camera stations, two of which overlooked Tonquin Pass. From these stations Campbell was able to connect with the 1917 survey of Yellowhead Pass, which extended south to this area. Altogether, Campbell surveyed twenty-two stations from Whirlpool Pass to Tonquin Pass. On September 9, Campbell and his crew were en route to Jasper when they met Cautley's crew.

Cautley started surveying the next morning. Platz began searching for gravel, and the crew started running trial lines. Cautley wrote: "I find that the information is incorrect and that Tonquin is much more of a pass than I had been led to suppose." After a rest day on September 11, "Mr. R & self went over entire pass and practically located all monuments, of which there will be six, besides two bolts, whereas I had been told four and made my arrangements accordingly. Observed in P.M." Surveying continued on September 13, and Cautley sent the packer back to Jasper for more food and monument supplies. Wheeler arrived that afternoon.

After Wheeler and Stelfox left Cautley at Whirl-

pool Pass on August 24, they reached Fraser Pass that evening and camped at a site north of the pass that they had used in 1920. In his diary for August 25, Wheeler wrote: "While at breakfast Jimmy Lamb came up after horses. Said Campbell was moving. Packed up and moved down valley. Arrived at Campbell's camp just as they were pulling out. Commenced to rain soon after." The men travelled for about 8 kilometres and stopped at a comfortable camp in a meadow. From there, on August 26, Wheeler accompanied Campbell and Archibald to one station, while Moffatt and some crew members occupied a station across the Fraser River.

The next morning Wheeler and Stelfox, accompanied by Lamb, started for Lucerne. When they stopped to camp that night, they discovered that "Pinto Frank had snagged himself in front leg at first joint." The next morning "Pinto Frank very stiff. Tried to leave him behind but he refused to be left and followed along, keeping well up." The men reached Lucerne at 1:00 pm and then started travelling to Jasper, camping that night at Dominion Prairie. On August 29, they reached Jasper in the early afternoon, and Wheeler left for Edmonton on the evening train. After spending a day in Calgary, he arrived in Banff on August 31.

Wheeler was there until September 9, when he took the train back to Jasper, arriving the following morning. Campbell reached Jasper in the afternoon. The two men had supper at the train station and talked about the surveying. Campbell's crew was camped at the Miette campground, but Wheeler stayed at a hotel because he had a cold.

The Boundary Commission report described Campbell's final surveying project for 1921.

In accordance with an arrangement between the Commissioners, Mr. Campbell with two assistants now proceeded up the valley of Miette River to its head in order to ascertain how many passes of the Continental Divide, lying between Yellowhead Pass and Robson Pass, would be likely to require survey and monumenting; also to ascertain the best and easiest line of travel to keep in touch with the watershed and continue the survey of it north of Yellow-

Campbell crossing alplands near Grant Brook on exploration trip. 89.03.35, Jasper Yellowhead Museum & Archives

head Pass next season. The exploration took until September 21st, on which date the party arrived back at Jasper.

Meanwhile, one man had been left with horses at Dominion Prairie in the Miette Valley, where there was an abundance of good pasture, and, on completion of the Miette exploration, Mr. Campbell and his assistant with the Packers and outfit returned to Banff by trail over Wilcox Pass, reaching that place on October 4th.

Even though he still had a cold, Wheeler left for Tonquin Pass on September 12, accompanied by Stelfox. Before departing, he made final arrangements with Campbell. In his diary, Wheeler wrote: "Got away at 10:30. Crossed Marmot Pass and made camp well up at Portal Creek at Campbell's camp. Very snug camp. Jack got supper very quickly. Feeling very sick and grumpy. Turned in immediately after supper."

Wheeler arrived at Cautley's camp around 2:00 pm the next afternoon, and the two men examined the proposed stations near the centre of the pass. Wheeler described September 14 as a "very glorious day—not a cloud in the sky—bright and sunny—just like summer." He and Cautley spent the day covering the entire pass and selecting the location of the monuments and bolts. Wheeler took one of Cautley's crew to assist him in placing the bolts. This included one at the summit of Tonquin Hill, drilled into a big rock. The next day

SURVEYING THE 120TH MERIDIAN AND THE GREAT DIVIDE

was warm and sunny again. Wheeler finished placing the bolts and building cairns for those stations. Cautley spent the day surveying at several stations, Robertson and some men chained distances, Platz worked on monuments, and the packers hauled gravel and cement. After a busy day, Wheeler had supper with Cautley and went to bed early. He and Stelfox left the next morning.

From September 16 to 19, Cautley and his crew surveyed Tonquin Pass. On September 20, as the work neared completion, weather conditions deteriorated. Cautley wrote in his diary:

> Very cold wind with snow flurries all day, turning to blizzard towards evening. Self read at 8-R and 10-R, but unable to read at 1-R and 3-R because snow prevented visibility. Mr. R. final chained and check chained between 8-R and 10-R and 1-R to 3-R. Mark built 3-R & 10-R. Packers collecting stuff. Took some poor photographs.

The next morning there was "snow with recurrent snow, snowstorm all day and hell of a wind," Cautley wrote. He surveyed at the two stations that he had been unable to read the previous day. Then the crew started the return trip, travelling about 20 kilometres to the camp at Portal Creek where they had stayed on the way in. There was about 50 centimetres of snow in the drifts on Maccarib Pass. Cautley noted that he left the stations at 3-R and 10-R "in old wooden forms to protect the very green concrete from frost." By September 23, Cautley and his crew were back in Jasper.

In describing Tonquin Pass, the Boundary Commission report stated: "Except as a scenic route and a resort for mountain lovers, Tonquin Pass will never be a useful one.… The best that can be expected here is a possible trail for tourist travel, and even that will be a large undertaking either by the Whistler Creek route or by the Portal Creek route." The report noted the unusual geography of the pass: "Tonquin Pass is, perhaps, more unique than any of the others. It has two distinct summits, which are just one mile [1.6 kilometres] apart and are separated from one another by an isolated ridge named Tonquin Hill.… Tonquin Hill is an isolated, mile-long ridge, rising to a definite summit at its northwesterly extremity."

Tonquin Pass was the final pass to be surveyed in the second section of the boundary survey. The report commented:

> Tonquin Pass is the eighteenth major pass of the Rockies to have been surveyed by your Commissioners and, in looking back, it seems extraordinary that no two of them can be said to resemble each other very closely, for, while there are points of similarity between some of them, there is not one that does not possess some striking characteristic of its own.

The boundary survey in 1921 was very successful, especially considering there were three parties involved. They needed to co-ordinate their work, and each was travelling in a remote area with no ready means of communicating with the others. For Wheeler, it was a hectic summer, compounded by the Stone climbing accident.

The fieldwork for the second section of the Alberta-BC boundary survey along the Great Divide was now completed. But there were still the survey calculations, reports, maps and photographs to finish, and these needed to be completed as soon as possible, so that the commission could publish a volume on this section of the boundary survey.

The surveyors also needed to begin planning for the third and final section of the project, which would take place in even more remote areas for both Wheeler and Cautley. From October 11 to 14, representatives of the federal and two provincial governments, along with Wheeler and Cautley, met in Edmonton. The boundary survey had gone on for nine years, and it had become imperative to complete the work as efficiently and expeditiously as possible.

The representatives at the conference believed that it would take three seasons for Wheeler's crew to complete surveying along the Great Divide up to the intersection with the 120th meridian. Cautley's surveying of the 120th meridian south would not take more than two seasons, so he could spend one more season extending the line north from where he had finished in 1919. In 1922, his work would "include the monumenting and final survey of the boundary established during 1920," along with continuing to work southward.

The representatives made some changes to Cautley's survey methods once he crossed the Wapiti River. They directed him to use

> discretion… in the production of the line according to the nature of the country encountered. All expediences contributing to speed consistent with accuracy are to be adopted, such as avoiding cutting ground logs, stumps, etc., or producing azimuth from ridge to ridge with inexpensive line-cutting between.

A change was also made in the monuments used on the line heading south, where

> permanent concrete monuments are to be established at intervals of approximately 6 miles [9.6 kilometres], subject to the Commissioner's discretion as to suitability of location. Standard Dominion Land Survey posts are to be established at intervening points of approximately one mile [1.6 kilometres] apart.

The conference representatives addressed Wheeler's extended absences, and declared that they were "of the opinion that Mr. Wheeler's supervision of the work is such that the efficiency of the work and progress is not materially handicapped by his absence."

The boundary line through Phillipps Pass, an issue outstanding since 1914, was finally resolved, and Cautley was instructed to complete this work at an opportune time during the next three years. With a plan to complete the initial Alberta-BC boundary survey, Cautley and Wheeler were ready to start the final section in 1922.

Cautley took this panorama from Tonquin Pass. e011205449, e011205450, e011205451, Library and Archives Canada

S-4332

S-4333

Felix Plante (centre), Percy Barraclough (left), Lloyd Shovar (right). Plante was born in Lac Ste. Anne in 1893, and his family was part of the Métis community that lived there. The family moved to Entrance, east of Jasper, and trapped in the area to the north. In 1914 Plante started working for Fred Brewster, a famous guide and outfitter in Jasper, and he gained a reputation as an outstanding guide. Plante purchased his own outfit in 1929 and operated a guiding business for over 25 years. He died in 1994 at the age of one hundred. e011166440, Library and Archives Canada

1922

In 1922, the British Columbia crew started surveying the most remote section along the Great Divide. North of Yellowhead Pass, there were no communities where they could purchase supplies and very limited communication possibilities. There were almost no previous surveys to connect with the boundary project, so to assist Wheeler, Deville made arrangements for the Dominion's Geodetic Survey department to become involved. This department sent H.F. Lambart to develop a triangulation network from Yellowhead Pass to the 120th meridian by establishing a series of stations that the British Columbia crew could connect to their surveying.

Beginning at a baseline at Yellowhead Lake that would connect with existing surveys in that vicinity, Lambart would set intervisible stations marked with large cairns several kilometres apart at locations near the boundary up to the 120th meridian, which was assumed to be in the vicinity of Jarvis Pass. Around that location, a baseline for the northern terminus of his triangulation network would be established.

Lambart worked for all three seasons of the final section of the boundary survey. In 1922, according to the Boundary Commission report, Lambart's "triangulation work of the past season was chiefly reconnaissance, consisting of the selection of suitable stations for the quadrilaterals of the triangulation, also the securing of necessary information for the measurement of a base on Yellowhead Lake the following winter." During the following two seasons, Lambart surveyed from the stations he established in the first year.

Lambart added a new type of technology to the boundary survey—the airplane. The results produced the first aerial pictures of the northern Canadian Rockies. The work with the boundary survey was Lambart's first experience in the Rockies, and he kept a detailed diary that contains many lyrical descriptions of the scenery. For all three summers, Felix Plante was Lambart's guide and head packer. Lambart's diary contains many references to Plante, providing more information about the early years of this famous Jasper guide.

Along the 120th meridian, Cautley's work had two objectives. He was to extend the survey south into the mountainous terrain and also to complete the survey work on the section the Alberta crew had cleared in 1920. Cautley would also incorporate a new element into his survey—a winter trip to the 120th meridian.

A.O. WHEELER

In the spring of 1922, Wheeler was busy finishing his work for the second section of the boundary survey. There were reports to write, maps to finish, photos to label and many other details to attend to, so Wheeler sent the British Columbia crew out with Campbell. Campbell started surveying north of Yellowhead Pass where Wheeler had finished in 1917. Since he had spent time in September 1921 making a reconnaissance trip up the Miette valley, Campbell was familiar with the terrain to be surveyed at the beginning of the season. From July 9 to 25, he surveyed up to Miette Pass, occupying eight new stations, and incorporating three stations established the previous season.

On the morning of July 12 Wheeler came to Jasper, where Colonel S.M. Rogers, the superintendent for Jasper National Park, met him. Then he went to Fred Brewster's place at Jasper Lodge, where he met Lambart and "took up quarters with him to the cabin." Afterwards Colonel Rogers drove Mrs. Rogers, Lambart and Wheeler to the airfield near Henry House, east of Jasper. In his diary, Wheeler wrote that the weather was "cloudy and windy, looked stormy. Col. & Mrs. Rogers went back to town. Lambart & I stayed and had lunch with the flying crowd." In the afternoon, Wheeler went

Flights over the Northern Canadian Rockies

In 1921, while surveying for the Geodetic Survey in the southern interior region of BC, Lambart first used an airplane for aerial reconnaissance. He found that an aerial flight over the region to be surveyed made it possible to select the stations for a triangulation network more quickly and efficiently. Lambart believed that in mountainous terrain an aerial flight would provide even more efficiency and cost savings. He convinced Deville to make arrangements with the Royal Canadian Air Force to provide a plane for a series of flights over the area along the Great Divide between Yellowhead and Jarvis passes. Captain J.H. Tudhope (a member of the Canadian Aviation Hall of Fame), along with a mechanic, flew a D.H. 4-B biplane from High River to an airfield near Henry House, east of Jasper.

The Boundary Commission report described Wheeler's thirty-five-minute flight on the afternoon of July 12: "It was a first experience and, to one who had been making maps for the best part of a lifetime it was a revelation to gaze from so exalted a viewpoint upon Nature's map as it unfolded beneath in a slowly moving panorama, for such was the impression received although the viewpoint itself was travelling at a speed of eighty miles [128 kilometres] an hour."

Wheeler and Lambart co-authored an article in the *Canadian Alpine Journal* describing the flights they took with Captain Tudhope. The next day, Wheeler made a longer and more adventurous flight. Tudhope piloted the biplane up the Miette River valley, where they saw Campbell's camp. From Miette Pass, the flight proceeded along the Continental Divide over Moose Pass, where they viewed nearby Mount Robson. "We were flying at 6000 feet [1830 metres]," Wheeler wrote, "and the clouds were not far above." Once they passed Mount Bess, "this was the end of my known country." Tudhope continued flying northwest along the divide past Avalanche Pass. "There was a high ridge of mountains to be crossed and the clouds were low over them. It was too dangerous to attempt crossing in the clouds, so of necessity it was a matter of beating around. This we tried to do, but got forced to the west, and eventually came out over a broad valley with a large river winding through it." Tudhope realized that he was in the Fraser River valley. An attempt to return to the divide was unsuccessful. "It was necessary to make a quick drop below the cloud belt. We swooped on a downward curve, the mist thinned and showed sheer rock walls all around. We were in a pocket enclosed by rock masses and roofed by a cloud bank.... Swooping out of the pocket, we followed the narrow, inhospitable valley of the West Branch of the Beaver River." Then the plane emerged into the Fraser valley again and they saw a village below. Tudhope landed in a field about 1.5 kilometres away. People soon arrived and the two men learned that they were at McBride, and were the first airplane to land at the village. More importantly, they were able to purchase gasoline, for the plane had insufficient fuel for the 200-kilometre return trip. The direct trip back to Henry House was uneventful.

A.O. Wheeler ready for his first plane ride. e011166443, Library and Archives Canada

In reflecting on his experience, Wheeler wrote:

Somehow I felt older, and that I had experienced one of the biggest things in my life. I realized the relaxation of an unconscious strain which had left me with a feeling of physical exhaustion, and yet to fly is to want to fly again....

As a means for mountain reconnaissance the airplane offers exceptional facilities. Given a clear day and the ability to keep to known landmarks, it is to a topographer a study of a living map, the most accurate that can possibly be had. I was enabled to get a clear conception of the country my future surveys would cover, and the nature of the access to them, and in one case was able to obtain information that will prevent a considerable loss of time. Nearly two-thirds of the desired ground was covered, and what I saw enabled me to secure a very fair estimate of the rest.

In conclusion I may say that the airplane service can be applied to distant and difficult survey work with great advantage, and I should think could be used as a means of saving expense.

On July 14, Lambart made his first flight, a trip that lasted about 1½ hours. Tudhope flew to the Moose Pass area, where he had to turn back. In his report, Tudhope wrote: "To avoid repetition of the flight with Mr. Wheeler, on account of the low threatening clouds, flight was abandoned after Moose Pass was reached. In the neighborhood of Mount Robson, Mr. Lambart checked up several of his survey points that he was in doubt of." (He had already established some stations for his triangulation network.) Lambart took fourteen photographs. He saw Campbell's camp in Miette Pass but did not see the crew. Inclement weather prevented flying the next day.

July 16 was a special day and Lambart wrote a detailed description of his flight.

Magnificent clear day—what we have been praying for. Leave the field at 8:53 and go

A view of Mount Robson from the north. Coleman Glacier is in the foreground and Snowbird Pass is in the centre. e011166445, Library and Archives Canada

Mount Bess, foreground, and Mount Chown. e011166446, Library and Archives Canada

by the Miette over Mons [one of Lambart's survey stations], the Colonel, up the Moose. Campbell had moved his camp and our camp was easily picked out at the mouth of Terrace Creek, a little spin to indicate that they had been spotted and on past Mt. Bess. Robson in all its glory passed on the left. 24 exposures made. A blanket of smoke confronts us north of Bess but fades away as we go through it.

The country north of Bess much lower, valleys wider right up to Mt. Sir Alexander and Ida. Flying almost along the watershed and a little west of it parallel to the Jackpine Pass. Alexander on the east, surrounded with immense ice fields. Fly towards Jarvis Lake and turn back just east of Mt. Ida, going right over the top of it on the way.

Back almost the same way till the summit of Moose Pass where Tudhope circled down into the valley of the Moose and passing over my camp dropped a good-sized mail bag.

Mount Sir Alexander. e011205423, Library and Archives Canada

From there, Tudhope flew down the Moose River to the Fraser River valley, up to Yellowhead Pass and back to the landing field. The flight lasted three hours and forty-three minutes and covered about 500 kilometres. Lambart concluded his account by writing: "This flight is a memorable one never to be forgotten."

Lambart attempted to take one more flight the following day. Tudhope reported: "Mr. Lambart wished to make a second flight over the same course, and it was intended to make an early start. Unfortunately, turning around at the end of the field to take off, the splice on the tail skid shock absorber pulled out, and we were delayed an hour and a half." By the time repairs were finished, smoke had started obscuring visibility. Tudhope and Lambart flew anyway but by the time they reached Yellowhead Pass, "the view becomes almost blocked out and we have little more to go by than the bottom of the valley and the mountains on the east of the Miette," Lambart wrote. The men returned to the airfield after a flight of thirty-five minutes.

In the *Canadian Alpine Journal*, Lambart emphasized the benefit of the flights for his surveying: "In connection with the laying out of

a triangulation net and assisting in the development of a topographical map, these flights proved of great value. The operations afforded the unique opportunity of seeing three separate mountain groups distinct from their surroundings, higher and heavily glaciated."

The airplane flights Lambart and Wheeler made showed that this technology could benefit surveying. Both men noted that in one day of flying they could plan the stations to be established for a whole season of surveying. Lambart's thirty-eight photographs were the first aerial photographs taken of the northern Canadian Rockies and provide an excellent record of the mountains as they were almost a century ago.

on a brief flight, and the next day, he took a more extended journey. After these flights, Wheeler returned to Banff and did not join the BC crew until mid-August.

While surveying in the Miette Pass area, one of the BC crew members, Alex Nelles, had to be brought into Lucerne, where he was treated by Dr. O'Hagan. (In 1922, Dr. Thomas O'Hagan was the physician at Lucerne, the divisional point for the Canadian Northern Railway west of Yellowhead Pass. Two years later, he moved to Jasper, where he was a well-known physician for almost thirty years.) This occurred while Wheeler was in Jasper, and his diary for July 12 stated: "Campbell arrived on evening train. Nelles very sick—ptomaine poisoning."

From Miette Pass, Campbell and the BC crew surveyed two small passes to the northwest, Grant and Colonel. Between July 26 and August 6, two stations established in 1921 were reoccupied, and nine new ones were surveyed. From the second pass, the crew proceeded down Colonel Creek to Moose River and up to Moose Pass along the divide. Between August 9 and 17, Campbell surveyed eight stations in the vicinity of Moose Pass. Then the crew moved across the pass and down Calumet Creek, on the Alberta side of the divide, surveying three stations in this area.

While they were there, Wheeler arrived on August 19. In his diary, he wrote: "My horse leaped at a creek crossing and tore my breeches all to bits—did not get hurt." Campbell and the crew were out surveying, but packer Jimmy Lamb "helped set up my tent—got comfortably settled." The crew had a long day and did not return until 8:00 pm. Wheeler continued: "Treated all round to cigars. Chatted with Campbell and went to bed."

From Calumet Creek, the survey crew bypassed the Robson Pass area. In 1911, the Alpine Club of Canada organized an expedition to the Mount Robson area led by Wheeler. With the assistance of the federal and BC governments, Wheeler made a phototopographic survey of Mount Robson and its vicinity and produced a map. The results of the expedition provided the impetus for the creation of Mount Robson Provincial Park in 1913. Wheeler's 1911 stations and photographs near the boundary in the Mount Robson area could be tied into the Alberta-BC boundary survey, avoiding dupli-

Station around Moose Pass. v771_pd_43_231, Whyte Museum of the Canadian Rockies

cation. Initially the men moved down the Smoky River to the junction of Chown Creek. Wheeler and Campbell occupied three stations in the area and then, on August 26, moved to Bess Pass.

On August 27, Wheeler and packer Leo Simpson headed north, trying to find Lambart. They met him the following afternoon. Lambart and his crew had completed their fieldwork and were on their return journey, so Wheeler and Simpson accompanied him back to Campbell's camp. That evening Lambart wrote in his diary: "A.J. Campbell is encamped on the Bess Summit and had been jumped from the head of the Calumet as Mr. W.'s photographs of 1911 through the Robson district would be used and needed no supplementing." Wheeler and Lambart's crew arrived at Bess Pass at noon on August 30.

Campbell had continued surveying between August 27 and 30, but because of smoke in the air, he was able to complete only one station. On the last day of the month, Campbell and Wheeler each occupied a station, but the smoky sky prevented good results.

Wheeler moved camp back to the Smoky River near the mouth of Carcajou Creek on September 1,

Smoky River (left and foreground), Mural Glacier (right), Mount Robson area (back left). v771_pd_43_247, Whyte Museum of the Canadian Rockies

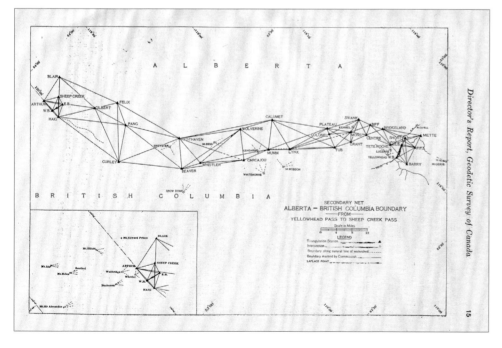

Lambart's triangulation network. e011166442, Library and Archives Canada

while Campbell and a couple of men camped at a location along this creek. The next day the air was finally clear. Wheeler surveyed from a nearby ridge, while Campbell surveyed from Lambart's geodetic station, Wolverine, where he made a connection with one of the boundary survey stations.

On September 4, the BC crew moved up Carcajou Creek, but the next day the clouds were over the mountains, and it started to snow. In the afternoon, Wheeler and Campbell went with packer Jimmy Lamb to explore the area. During the following days, snow continued, so in the late morning of September 7, Wheeler abandoned fieldwork for the season, and the crew moved to the mouth of Calumet Creek, where they remained for a day because of snow.

On September 9, it warmed up and rained, so the men rode out, following the trail past Mount Robson and down to Robson Station on the rail line. From there, the crew proceeded to Yellowhead Pass and they arrived in Jasper on September 13. In his diary entry for September 12, Wheeler wrote:

> Packed up. With Nelles pushed on ahead of pack train. Going across Grant Brook Bill fell in middle of stream and I got a ducking. Had to take off all my clothes and wring them out. Put them on again wet and continued travelling. Called in at Lucerne about 2 PM and paid Mrs. Weaver Nelles' hospital bill of $35.00; also called on Dr. O'Hagan. He was away. Paid Mrs. O'Hagan $25 on Nelles' account. [Mrs. Weaver had a nursing residence next to Dr. O'Hagan.]

Smoke from forest fires limited the surveying that Wheeler and Campbell were able to complete in 1922. The BC crew covered less area than had originally been intended, but there was a compensating factor, as Lambart noted: "The photographic work has been very hard, and if it hadn't been for the photographs taken by Mr. W. in 1911 very greatly supplementing the work they would have been very badly off for next winter's plotting operations."

H.F. Lambart

Lambart arrived in Banff on the evening of June 1 and spent the next day with Wheeler discussing the surveying for the field season. In his diary, he wrote: "Mr. Wheeler is very lucid and clear in his way of descriptions and I got a very good first-hand knowledge of what was in front of us this summer." Lambart also "wired Brewster for pack train, grub list completed and sent to Macdonald & Cooper, Edmonton." Then he left for Vancouver for a few days, and arrived in Jasper on June 10.

An article in the *Edmonton Morning Bulletin* on June 17 outlined Lambart's work:

> The federal government is co-operating with the provinces in the work of determining this [boundary] line, and to this end have detailed H.F. Lambart from Ottawa to assist the boundary commissioner, by taking charge of the triangulation control....
>
> Further aid will also be obtained from the federal authorities. This will take the form of assisting the surveyor to make a preliminary investigation from the air.... For this purpose a machine will be allocated for duty from the air station at High River. The aircraft will carry surveyors over areas that are now little known and reveal topographical features of importance to the work in hand. It is expected this reconnaissance will facilitate matters and appreciably hasten the completion of the survey.

Lambart spent five days getting organized at Jasper. His surveying assistant was Percy "Barry" Barraclough, an engineer and surveyor who had been on the BC-Alaska boundary survey and had spent four years as a soldier in World War I. His packer and guide was Felix Plante, while Lloyd "Slim" Shovar was wrangler. As part of his equipment, Lambart took a portable phone set, hoping he would be able to use it to communicate when he was along the national park's telephone line.

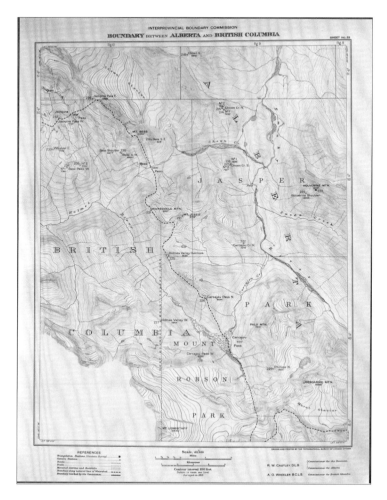

This map covers the area north of Mount Robson. "Boundary between Alberta and British Columbia" map atlas, Sheet 33, Land Title and Survey Authority of British Columbia

Lambart and his crew departed on June 15.

> Got away at 12:30, made Geikie Station at 7. Crossed right at the mouth of Miette and followed the left valley of the Miette by trail and finally got on the old CNR track & we followed it most of the way into Geikie. The bridges are out & even when in we had to go up the valley & climb up the other side. We had one bad crossing where we nearly lost a horse and before that one rolled down the embankment through large boulders and I thought for sure Brewster was minus a horse.... It is

really a crime to see this old C.N. railroad grade abandoned and going to decay with the ties still rotting in their place. This line was only in operation 5 years about when the rails were shipped to France.

The next day the crew reached Yellowhead Pass, and a freight train dropped off their supplies. Lambart wrote that he was "very disappointed, no results with telephone." On June 17, Plante and Barraclough investigated the trail ahead, while Lambart was at the ranger's cabin doing office work. He also tried "out the telephone on abandoned poles along the line and a ditch full of water for ground. No success." While Plante continued working on the trail the next day, Lambart remained in camp: "Telephone changed this time trying an aerial between the Ranger's flag pole and also trying two different grounds with no success."

On June 19, the crew made their first climb, going up Tête Roche. Lambart wrote:

> Follow Campbell's trail of the autumn of 1921 over the ridge leading up the valley of Miette River. On reaching the summit of the escarpment turned west off the trail and from this point to the base of Mt. Tête Roche I never passed through such beautiful alpine flowers, simply a sea of bloom…. We swung around to the west side and made a very easy ascent. The face of the mountain in front of us facing the Miette is a sheer cliff.

From the summit, Lambart had his first view of Mount Robson. Below them was Lucerne, and he and Barry made their descent in that direction, winding up in "terrible windfalls." At Lucerne, "Mr. Young kindly took us back to camp in his speeder getting in about 7."

The next day Plante and Shovar took a load of supplies up the Miette River valley, while Lambart and Barry walked up Miette Hill. On June 21, the two men established the two stations to be used for the baseline that would be established at Yellowhead Lake during the upcoming winter. "This we did successfully and got a splendid double expansion from Lake Lucerne

[Yellowhead Lake] to both sides of the valley." During the day, Plante "went to the second crossing of the Miette and swapped with Brewster a good horse for the half gelding that rolled down the embankment and has been giving us trouble." Rain prevented the crew from departing on June 22. The telephone, which never worked, was packed up and sent to Calgary. "The aerials were left standing and the ground wire with the broken alpenstock was left in Greentree's shack," Lambart wrote.

On the morning of June 23, Plante and Shovar started up the Miette River valley, while Lambart and Barry went back to Yellowhead Lake. There they put in a signal for the east base and cleared out some of the brush. Early in the afternoon, the two men followed Plante and Shovar, arriving at the camp they set up close to Miette Pass. Lambart described the scene: "We are here in a most magnificent open part of the pass on a little dry knoll, much snow in patches all about, flanked by trees and spread out before us the beautiful Miette range along the whole western side of the valley. The east side is cut off by a central rounded hill on which we have our station Centre."

The next morning, Lambart and Barry established station Centre and then explored the area. June 25 was "a Sunday in camp in our beautiful surroundings. Get my table out in the open and typing a great part of the day, on descriptions to date, etc." In the late afternoon, smoke filled the valley, a harbinger of their main difficulty for the summer.

The following day, Shovar remained at the main camp while the other three members went on a fly camp to Mount Bridgland. In the afternoon, Lambart and Barry established a geodetic station on the mountain, while Plante went hunting and got a caribou. On June 27, the three men moved up the valley, picking up the caribou meat along the way. They went to a high bench above Miette Pass and camped at the edge of the timberline. It was cloudy in the morning but cleared in the afternoon, so Lambart and Barry established their next station at a high location. Unfortunately, they couldn't see their station on Bridgland, and later another station had to be established there.

The men then moved northwest into the Grant Brook drainage. Lambart wrote:

We passed through magnificent meadows simply covered with the western anemone and passing up through an open gulch we came upon four caribou in the open who came towards us at first and eyed us curiously and with their noses turned up sniffing the breeze. They stared at us until they finally got our scent and off they went.… We traversed the wide basin of the north branch of Grant Creek and passed into the narrow gorge of the west fork of the Grant keeping up on the north side of the valley high up.… One horse went over backwards over the slippery shale slide and was recovered fortunately without damage to pack or horse.

He described their camp site.

We made camp on an elevated knob which commanded a splendid view of the valley ahead and apparently not bad feeding ground in sight. The Colonel off in the distance at the end of the valley. We are surrounded very closely by the mountains and now with small glaciers more numerous in sight.

Smoke filled the air on June 29, so Lambart and Barry explored the area, while Plante and Shovar went back to their previous camp to bring up supplies. Lambart and Barry established a station above Grant Pass on the last day of the month.

On the first day of July, "the mountains cut off with smoke worse than they have ever been. Climbing out of the question and Barry and I set out up the valley and explore[d] the country to the head of the Colonel Pass. A very delightful walk and a joy to see the magnificent country." July 2 was a Sunday in camp. "It is very magnificent amongst the heather and open, commanding a view of the pass and the mountains east and west."

On July 3, it rained early in the morning and again late in the afternoon. In between, the surveyors established station Grant. Twice they had to stop and light a fire. "Proceeded to build the 6 foot [1.8 metre] cairn

with pole and flags—having arrived at 1:30. We stayed on top until 3:55 hoping to see something but was disappointed, only getting sight of one point—Centre." While they were working, the packers moved a load of supplies to their next camp. Lambart concluded his diary entry for the day: "Heavy rain and lightning as I write (8 P.M.). It's bully being in a comfortable tent and warm in such weather."

Good weather followed for the next two days. The crew crossed Grant Pass and proceeded along the Alberta side of the divide towards Colonel Pass. Before leaving the Grant Pass area, they established one more station. Their next camp was on a small lake at the headwaters of the Snaring River, where they stayed for four nights. They established a station northwest of the camp that was about 900 metres higher and provided an excellent view of the terrain ahead. Lambart saw Mount Robson again, and in the far distance, he viewed Mount Sir Alexander for the first time. He and Barry spent almost four hours establishing the station and examining the landscape.

After two days of climbing, Lambart and Barry rested for a day. North of Colonel Pass there was no feasible route along the divide. The existing, seldom-used trail followed Colonel Creek down to the Moose River. Lambart sent Plante and Shovar to clear the trail ahead and establish a camp at the junction of Colonel Creek and Moose River. The men did not return until the second day because of the terrible condition of the trail. Lambart wrote: "I don't take very kindly to these unforeseen delays. We have one satisfaction, however, that it is there for the use of the next man that comes along, who I guess will be Campbell and possibly Mr. Wheeler."

On July 8, Lambart's crew travelled across Colonel Pass and down Colonel Creek to Moose River. Lambart described the view from the pass.

Straight down the valley Robson loomed up as brilliant as a shaft of silver against a slightly leaden sky. White frost last night had left the air very brilliant, it was the natural framing which struck one so forcibly. On the right the Colonel and on the left the opposite side of the steep valley of Colonel Creek, the foreground the dark outline of the mountains on the opposite side of the Moose and in front the trees.

The trail was difficult, and when Lambart finally reached Moose River, "I saw no way of crossing it and as the trail up the Moose lay on the other side I decided we should make a raft and when the train came in to raft the stuff across. So after lunch Percy [Barraclough] and I got busy and got all the timber together on a skidway, but when the train came in Felix told us of a ford a little way above." Lambart had a mishap. "Very unfortunately lost my second pair of glasses for this year on crossing the Moose. As I was getting into the saddle the end of the gun in the case flew up, breaking and knocking them into the water."

On July 9, Plante went back to bring the rest of the supplies from the previous camp. Lambart and Barry constructed a raft, since they needed it to reach a station across the river from their camp. The following day they established this station. En route, they saw a band of twenty-four mountain goats. "Just after reaching the summit the clouds closed in entirely and we had snow, but after we had finished the cairn we hung about a while hoping for a glimpse but none was to be afforded."

Lambart left for Jasper with Plante and Shovar on July 11. They travelled on a good trail down Moose River to Moose Lake, then west to the train station at Rainbow. Plante returned to camp while the other two men flagged a freight train to Lucerne. That evening they took a passenger train to Jasper.

Lambart returned from his air flights on July 20. "Leave Jasper by the 9 AM to Rainbow…. My glasses in the morning mail just before leaving. Phone taken minus receivers which had not arrived from High River." At Rainbow, Lambart met Plante, who had been waiting there for three days. After lunch, the two men travelled up the Moose River valley, reaching the new camp at Terrace Creek by 8:50 pm. While Lambart was away, the crew moved camp and established a cache at Moose Pass.

Dense smoke and rain prevented work the next day. In his diary, Lambart wrote: "Many points about the valley in coming in last night with Felix I completely overlooked when flying over last Sunday. I attribute this to the fact that so much of our attention was focused on Robson that it was impossible to keep one's eyes off of it. The peaks down the valley of the Moose are really beautiful." It continued raining the following day. Lambart went hunting and exploring around Moose Pass, "returning drenched without seeing a moving thing." July 23 was Sunday. Lambart wrote in his diary: "Uneventful day except for Slim narrowly escaping a bad cut on the head from an axe, as it is Barry did some good work on a small cut and a pretty bad bruise. At last we have meat in camp. After fruitless efforts of all of us, Felix finally got a caribou."

On July 24, Lambart wrote, it was "two weeks since our last climb on Tub [geodetic station], today do Lynx." After crossing Moose Pass, the Alberta-BC boundary heads in a southerly direction for several kilometres before resuming its general northwest course. Lynx is located near the end of the southerly direction of the divide. Wheeler had established a phototopographic station on the mountain in 1911 but Lambart set his station lower. "The extra exertion did not seem warranted for the Geodetic Station as all the back points were visible, and the only loss we may sustain is the selection of the proper points ahead. However, I have a few to choose from." Plante accompanied Lambart and Barry up to Snowbird Pass and had tea ready for them when they returned. "This is the first occasion on which we have taken Felix and on which we have used a rope. Both went well & in fact without would have been hazardous, a few crevasses passed."

The next day, the crew moved camp over Moose Pass and partially down Calumet Creek. Lambart described the scene:

> This is a perfectly magnificent spot. Straight down the valley we have Mumm Mountain presenting a straight wall on the north face. Above the camp stretches one of the most beautiful alplands I have seen, a sheet of the most magnificent green and covered thickly with flowers…. Two small glaciers form the course of the Calumet which is roaring down the valley behind the tent as I write. Between these two glaciers stands

the beautifully banded peak, Calumet, which I hope to climb tomorrow.

Lambart also spotted Wheeler's 1911 cairn on Calumet.

Lambart and Barry attempted to climb Calumet on July 26 but stopped within 35 metres of the summit. The next day they returned and brought Plante with them. They made the south shoulder of Calumet about 300 metres below the summit. "This shoulder of the peak serves our purpose well and was all we wanted to see. Felix was along with a rope but we had no need of it." At the end of his entry for the day, Lambart wrote:

Heavy rain setting in in the evening which I trust will not prevent us from moving tomorrow. If the two points I have chosen ahead work out satisfactorily, I will be on the other side of the trouble zone, i.e. Mt. Bess and Chown. We are just leaving A.O.W.'s Mt. Robson sheet and any information we now get will have to be from Miss Mary L. Jobe. [Jobe, guided by Curly Phillips, had explored the area north of Mount Robson in 1914 and 1915 and produced the Jobe-Phillips map. Mount Jobe near the boundary is named for her.]

Lambart and the crew's next camp was at Berg Lake at the base of Mount Robson. His diary entry for July 28 began: "Here we are under the highest mountain we have in the Rockies and the 'show place' of the mountains. The country is plum full of smoke again so we cannot get the full benefit of all that surrounds us here." He noted "three teepees and a cabin nearby—evidently a packer's outfit." Lambart intended to stay at the lake only for a couple of days, but the dense smoke that had affected surveying all summer kept the crew at this camp for over two weeks.

Lambart wrote in his diary the next day: "Heavy smoke envelopes us again and a day lost, had hoped to do Mumm" (the first mountain along the divide northwest of Robson Pass). Lambart dispatched Plante to Jasper to send and pick up mail. July 30 was a Sunday and there was still smoke in the air, so the three men remained in camp. Plante returned in the evening with mail.

Lambart tried to establish a station on Mumm Peak on July 31. "Very sorry to have to come down from Mumm after doing 2400 feet [730 metres] on it, turned back by smoke obscuring all but the immediate ranges. Afternoon finishing my writing. The telephone tried for the last time. A bunch of tourists up from below and camped just beyond." The smoke continued on the first of August. The next day, "Slim had us routed out and eating breakfast at five. But I for one got back to the eiderdown as the weather is absolutely hopeless, the smoke denser than ever."

On August 4, Lambart wrote:

Eight years ago the Great War commenced. The boys have been getting pretty stale, this constant loafing about camp, and are off—Felix and Slim with their lunches down the Smoky, Percy with the 22 [rifle] in another direction. I am alone with my thoughts and books and content. Work is out of the question, the top of Robson just visible, and no more than four miles [6.4 kilometres] away. The afternoon spent in a very interesting study of a colony of beavers on Adolphus Lake.

While still waiting for the smoke to clear on August 7, Lambart, Barry and Plante took a trip up Robson Glacier. (Slim went into Jasper for the mail.) Lambart recorded the recession of the glacier: "Recover A.O. Wheeler's reference points of Aug. 10, 1911 on E. & W. side of nunatak marking position of forefoot of glacier. Last measured in 1913, since that date recession on east side 259 feet [79 metres], on west side 212 [65 metres]."

Lambart was becoming discouraged by the delay: "The immediate future success of the work looks very gloomy & I am at a loss exactly what to do if this keeps up. I would not like to change my present program." Two days later, Plante, Barry and Lambart rode over Robson Pass and down the Smoky River to Carcajou Creek. "The P.M. has been very smoky and also cloudy, and now it is raining. I am at a loss what to do about the work, we could move direct to Bess Pass and let Mumm go with just putting a cairn up, but then the

Crossing Mumm Glacier. e011166444, Library and Archives Canada

SURVEYING THE 120TH MERIDIAN AND THE GREAT DIVIDE

link is broken and I am at sea as to the connections ahead."

Lambart and Barry finally climbed Mumm Peak on August 11. Lambart had loaned much of his equipment to a group attempting to climb Mount Robson. "It is not an extraordinary climb but without ice axes or alpenstocks our difficulties were magnified at the top." The summit was in clouds, so they "went down to the first rock where we made tea, making fire with one of the wooden sticks we had as substitutes for alpenstocks. While there it cleared wonderfully and we hiked to the top again. I managed to go over the country ahead except that part west of Bess, and cut in all possible points."

That evening Lambart wrote: "I shan't venture on another climb without the axes, Putnam has them and this is the third day since they left to do Robson. I am rather anxious about their safety and also for my axes as I wish to pull out in the morning."

Now that he had finally established a station on Mumm Peak, Lambart was ready to move camp on August 12.

> Routed out of bed by 5:30 and breakfast over by 6 and commence packing up with glee, but it starts coming down in sheets. This does not dampen our ardor at getting away from this place, and we pack up and get out in the pouring rain. Down the Smoky we go with our fourteen [horses], but for me it is walking as the rain is too cold, and under these conditions sitting in the saddle is not all it cracked up to be. We went along without a hitch until 2:45 when we halted and made our camp just beyond the mouth of Pamm Creek on the banks of the Smoky. Steady pour all day and everything drenched. A blazing fire and some hot tea and some lunch made the camp go up in fine style.

The next day, Lambart got Wolverine, the second station that he wanted to establish before Mount Bess. "We were a long time on the top trying to get our points between the clouds." There were visitors waiting for Lambart when he returned to camp. "Found Mr. Putnam and his boy and Hargreaves in camp as happy as they could be over their climb of Robson; truly a remarkable feat for such inexperienced fellows. The old man was with them in their bivouac two days on the mountain."

On August 14, Lambart and the crew departed early for Bess Pass, along the Great Divide. "Percy and I move out ahead and enjoy the enchantment of seeing a new and beautiful country unfold before us as we ride." Lambart decided to camp at the pass and spent the afternoon exploring the area. The next day the crew travelled along the BC side of the divide, then crossed Jackpine Pass back into Alberta and proceeded down the Jackpine River, where they stopped to camp by 1:00 pm. Lambart described the location as "the finest valley we have been in, very wide on the bottom and the flanking mountains—with the exception of those at its head—beautifully rolling with hanging valleys and immense stretches of upper grass land." Lambart again went out to explore and found two stations that he could establish from their camp. That evening he wrote: "I feel we are in the cream of things now with a way solved out of our difficulties in getting past Chown and Bess without one stiff climb to accomplish this and plain sailing in beautiful country ahead. Weather permitting, we will make good progress before the work shuts down only two weeks ahead."

However, the next day Lambart was "at a standstill again, clouds right down in the valley where they stay and rain during the afternoon. But oh, really, at heart I am happy, what a cleaning up, sleep somewhat needed, descriptions to write up and this diary at last brought up to the last moment."

On August 17, they established a geodetic station that was less than two hours from camp. Lambart commented on the terrain: "These immense stretches of uplands are a new feature and seem characteristic of the Jackpine watershed." Plante shot a bighorn sheep "and the change in our fare is welcome." The following day, Lambart established the second station. Since it was a walk and not a climb, Shovar accompanied Lambart for the first time, along with Plante. It took five hours to reach the station, but "the day was one full of rich experiences and fresh worlds seen as the view from the summit of Beaver was splendid, giving me the first real sight of the country stretching to the NW and showing everything between ourselves and Kitchi and Ida which stood out against a leaden sky quite distinctly."

Lambart decided to establish a third station in the area, named Resthaven, climbing "the point to the north in the Chown ice fields. Elevation 9000 feet [2750 metres]. A magnificent point." Lambart could see several stations all the way back to Calumet, along with "new points to the NW." At the end of his diary entry for that day, he wrote:

> This climb, which took us about 4½ hours from camp, is my third major climb of the week besides three smaller ones, and I learnt this evening after staggering into camp at 9 and crawling into bed the limits of my physical endurance. But we have some great consolations on completing this point, as our battle royal for the summer is over in gaining a safe and successful scheme past the Mt. Bess and Mt. Chown group.

August 20 was "a very welcome day of rest and the skies properly give us their benediction in showers of rain all morning." Lambart developed his recent photographs and was disappointed to find that most of them were overexposed. He noted: "This week is our last of real toil in the field and I hope and pray for the chance to make a satisfactory winding up of the season's operations in establishing at least two more points, if not four."

The next day they travelled 5½ hours down the Jackpine River, camping in a little meadow, and by August 22, they were on the west branch of this river. Lambart began his August 23 entry:

> In four hours travelling we reached this magnificent spot near the very headwaters of one branch of the Jackpine north and west of Jones and Meadow passes, although we do not know absolutely as we failed to connect with any well-defined trail during the day.

He described the locale as "a series of terraces with richly covered flowers and grasses, hundreds of little streams, waterfalls and the whole divided by ribbons of trees and patches of the same in isolated clumps."

On August 24, it was a clear morning, and Shovar woke the men early. Lambart and Barry left by 7:10 am and reached the station Curly (named for Curly Phillips, a well-known Jasper guide). It was located on the southern portion of Lambart's geodetic net, not far from the Great Divide.

> The view was unobstructed in all directions since this is one of the high points of the low mountains lying between Mt. Alexander and Mt. Robson. I am trying to fix the next two points ahead which will not be occupied till next year. If I can fix these points well and identify them on the Phillips-Jobe map, this and the station I hope to get on Saturday will not need to be climbed next year in picking up the reconnaissance again. The location of this point has been very satisfactory to me, it's a good top. The whole country is visible and the orientation and cutting in of the points lying to the westward and elsewhere have been accurate and check with the cuts from Resthaven and Beaver.

On August 25, the crew reached the headwaters of the Little Smoky River, travelling over Big Shale Mountain. They stopped in the first timber "in a very old Indian camping ground strewn with many poles which Felix says are old meat drying racks."

The next day, Lambart and Barry went to establish their last station of the season, which they named Felix, on the north end of their triangulation net. The men left at 6:10 am, but by the time they arrived, at 2:00,

> the whole country is so full of smoke that not a single point is visible and at first it seemed as if all we could do was to build our cairn and go. The cairn and flag is established and lo, just before starting, it clears just enough to see Curly, Resthaven and

Felix Plante with pack train. e011205425, Library and Archives Canada

Chown. We quickly get the plane table into action and have the location of this point (Felix) fixed, together with several others I have cut in in this district from Beaver and Resthaven and Curly.

The men arrived back at camp at 10:30 pm. Since there was an open view from the Big Shale

Mountain area, Lambart wanted to establish one more station nearby before leaving. He hoped to do this on August 26, "but smoke covers everything," so he decided to wait one more day.

It was a Sunday in camp, and in his diary, Lambart summarized the season's work: "We are not more than 35 miles [56 kilometres] from Mt. Sir Alexander and Mt. Ida. I feel well pleased with our progress, and feel if it

Cautley's winter trip. A tent is visible in the trees on the right. e011205453, Library and Archives Canada

hadn't been for the two weeks lost at Robson Pass we would have been very near our summer's goal." Lambart noted that the men worked well together "without a disagreeable experience," and he made only positive comments about each of them. "Slim has turned out better than was at first expected and has made us very comfortable although without any frills." In describing Plante, he wrote: "I prize him highly and would like to have had him on all the expeditions I have been on where horses are to be cared for. He is gentle with the horses and takes good care of them. He is a good packer." Lambart wrote about Barry: "Barry, the stolid chap has done well and it has been a great pleasure to have had one of the Old Boys with me again."

The next morning, Lambart and Barry climbed to a point about 450 metres above camp. It took the men about an hour to reach the location, a station they named Pang. "It was the greatest satisfaction to me to have made this last climb as it gave me what I most wanted—two cuts on mountains cut in from Curly & therefore fixing another quad ahead. Besides, we fixed in position the place of both Ida and Mt. Alexander."

Now Lambart was ready to start the return trip: "Our journey back was punctuated by nothing out of the ordinary except that the horses seemingly knowing they were homeward bound had an unusual amount of pep and covered the ground quickly." The crew met Wheeler late in the first afternoon and he camped with them that night. "Three of us [Wheeler, Lambart and Barry] tonight in the tent before a blazing fire and enjoying one another's company." Wheeler and Simpson accompanied Lambart's crew back to Bess Pass, where they arrived at lunchtime on August 30. Then the rest of the crew headed down to the campsite at Pamm Creek while Lambart remained:

in response to a very pressing invitation from Mr. W. so that the boys could write and get out mail. I spent a very comfortable evening and night in the little tent with Mr. W, he and Campbell and I talking the summer's work over.... I enjoyed a bath in the creek and a sleep in the afternoon and a first class sleep on a good bed of boughs which Mr. Wheeler insisted.

The next morning, accompanied by one of Campbell's packers, Lambart rejoined his crew. The men reached Robson Pass that afternoon, and there they found Curly Phillips with Governor General Lord Byng and his staff camped in the vicinity.

Lambart remained at Robson Pass on September 1. "We plan to spend the day here to straighten out equipment and store what I can at Hargreave's cabin and clean up my descriptions and other work. His Excellency and party pass in the morning on their way down the Smoky, Curly Phillips leading his train of about 12 horses." The next day Lambart's crew reached Robson Station and the close of the field season.

R.W. Cautley

In 1922, Cautley resumed surveying the 120th meridian. He had two parts to his operations. First, he needed to survey the meridian line south from the Wapiti River (where he had stopped at the end of the 1920 field season) to the mountainous terrain. Then he had to construct monuments and survey the section covered in 1920. Cautley would need a large amount of food and supplies for his work. He thought that if he had some of his supplies brought in during the winter, when the ground was frozen and sleighs could be used, it would facilitate his work during the field season. Cautley received permission from the three governments to use part of his budget to have supplies delivered to the Wapiti River. He made arrangements with Alex McIntosh and Ewen MacDonell, who had a business in Rio Grande, the small community in Alberta that was closest to his surveying.

The Boundary Commission report described the situation.

The total weight of outfit required for the season's operations amounted to 40,000 lb. [18,144 kilograms], including 22,000 lb. [9980 kilograms] of cement, 14,000 lb. [6350 kilograms] of supplies and 4000 lb. [1815 kilograms] of camp outfit. Besides the above 40,000 lb. of general freight, it was also necessary to provide for the transportation of about 40,000 lb. of gravel that would be required for the construction of concrete monuments. As there are no roads of any kind along the line to be surveyed, or waterways that would permit water transportation, it meant that 80,000 lb. [36,288 kilograms] would have to be handled by the packtrain, apart from ordinary camp-moving work.

Under the above circumstances it was essential to the success of the season's work that as much freight as possible should be placed at convenient points on the line for the actual commencement of surveying operations. Otherwise the packtrain service would have been quite unable to keep pace with the progress of the survey. It was therefore decided to try to place four caches in close proximity to the line by winter freighting. Of the above four caches three were comparatively small, consisting of cement and other monument material only....

The most important cache which it was necessary to make was one of 24,000 lb. [10,866 kilograms], including most of the food supplies besides cement, on Wapiti River.

McIntosh agreed to make a trial trip if Cautley accompanied him, and to take a contract for delivery of the freight if a practical route could be located.

Cautley took the train to Grande Prairie on January 23, arriving in the evening. On January 25, he "left at noon by team with McIntosh who had come to Grande Prairie for me by arrangement. Beautiful day with only just enough snow for sleighing." They travelled about

Canvas boat. e011205455, Library and Archives Canada

30 kilometres, about half the distance to Rio Grande, before stopping at a residence for the night. The next day the temperature was colder and there was a blizzard. The men reached Rio Grande at dark.

While the blizzard continued on January 27, Cautley and McIntosh "got grub outfit together." It was still cold and snowing the next day. In a letter to Deville, Cautley wrote:

> We started for the river on the 28th during a five day snowstorm with one other man. The furthest point upstream at which it is possible to get down to the river from the plateau is… 11 miles [18 kilometres] east of Rio Grande. We traveled 37 miles [59 kilometres] up the river from Chase's… and three miles [4.8 kilometres] downstream from the crossing of the boundary.… We made very slow progress owing to the depth of snow, which made it necessary to break trail for the horses on snowshoes but fortunately [we] had a big, powerful team.

Cautley reached their campsite on February 2. Over the next nine days, the men built a log cache to store the supplies McIntosh would bring later in the winter. Cautley also spent time making a survey of the Wapiti River, the largest one in the southern section of the 120th meridian, since it would be easier to do on the frozen river. The Wapiti River makes a loop along the boundary line and intersects it twice. There was also some steep terrain, so Cautley wanted to map the area. The weather was cold, with the temperature reaching about -40°C. On February 12, the men started the return trip, arriving back in Rio Grande two days later. By February 17, Cautley was back in Edmonton.

Cautley's field season started on May 25 when he left Edmonton. His crew consisted of eighteen men and thirty-two horses. Cautley's assistant for the last three years of the boundary survey was A.O. Gorman. Gorman, an experienced surveyor, was hired so that Cautley could focus on the organization and supervision of the surveying. The crew spent a few days at Grande Prairie getting organized. Cautley made arrangements to have the men driven out to Rio Grande and the outfit taken by wagon. Unfortunately, the "motor got the men out but broke down and was unable to make a second trip with outfit coming in wagon. Outfit arrived in farmer's hay rack at 8 PM."

The crew left Rio Grande on June 2, travelling 25 kilometres. The next morning, the packers found that "all horses went back and were only recovered by 1:30 PM. One of the backs of the new horses came in in very bad shape and has practically lost his hindquarters. I think he must have been hit by a falling tree in the brulé or got mired and injured his back plunging." The men reached Calliou Lake that evening, but had to leave the injured horse along the trail. On June 4, the head packer and Cautley spent the day going about 30 kilometres back to retrieve three horses. Since the injured horse couldn't stand up, they shot him. The next day it snowed, so they remained at this camp one more day.

On June 6, the crew reached the Wapiti River crossing: "River too high to ford. Assembled canvas canoe and went across to cache which does not seem to have been disturbed and is very full. Made raft and got ropes across river." However, the next day, "self and some men open cache and started rearranging stuff, during which it was discovered that about 2100 pounds [952 kilograms] of it had been stolen. Harry and part of pack train was leaving for Rio Grande at noon and I went with them." When he arrived in Rio Grande on June 8, Cautley wrote to Deville, explaining the situation. The theft was reported to the Alberta Provincial Police, who sent an inspector from Grande Prairie to Rio Grande the same day. Cautley told Deville that "fortunately nothing except food supplies and tobacco was taken." He also commented: "I have succeeded in replacing the stolen goods from the wholesale houses in Grande Prairie by long distance phone and expect to leave for camp in the morning."

By June 10, Cautley was back at camp, and he and the crew spent the first three days surveying the meridian line and transporting supplies across the Wapiti River. The tall steep riverbanks and the high water of the river made the work difficult. As in previous seasons, heavy brulé made the work slow. The packers ferried the large amount of food and supplies along the line and moved camp. Mark Platz, in his fifth season as a monument-builder, worked at establishing the monuments. In his diary entry for June 23, Cautley wrote that he and Gorman went to station 65-5 at night and took three observations on the stars, their first of the season. Cautley also noted that Platz was in camp with a bad hand, likely a cut that got infected. In his diary entry four days later, Cautley wrote: "Lanced Mark's hand which is swelled 3 inches [7.5 centimetres], thick and full of pus. I took all kinds of care but am rather afraid of what might develop." Platz recovered without further complications.

Cautley's entry for July 1 noted heavy smoke in the air. As it was for Lambart and the BC crew, smoke was the main difficulty for surveying during the 1922 field season. The work proceeded steadily through July. The weather was good, with very little time lost because of rain. As the Alberta crew headed south, the elevation increased steadily. On July 18, Cautley wrote in his diary that he and a crew member climbed a hill near the line that was over 1800 metres in elevation: "We saw a lot in spite of the smoke and note that all the tributaries of the Crooked [Narraway] River seem to run in box canyons 400 to 700 feet [122 to 213 metres] deep."

Four days later, the Alberta crew cut the trail to the ford across the Narraway River. That evening, the federal inspector of surveys, A.M. Narraway, arrived and

Horses swimming Wapiti River. e011205454, Library and Archives Canada

stayed for three days. Cautley wrote a letter to Deville on July 24 for Narraway to send.

Since the 12th June the party has surveyed 27 miles [43 kilometres], most of which has been through heavy brulé on line and we are now near the south boundary of Tp. 62 and on the north bank of the Crooked River. Across the river the line runs up high on the westerly slopes of a widespread mountain, with a probable altitude of about 8500 feet [2590 metres] and a number of outlying ridges of considerable height. I consider that the above mountain marks the entrance of the 120th meridian into the true mountain area, and intend to end my season's work towards the south on the highest point of the mountain well above timberline—where I shall place the final bolt. I also intend to occupy a station on the summit of the mountain, and to erect a large cairn there as a point of identification for future work. The final bolt will occur in Tp. 61. I believe the above course to be in accordance with the express intention of the

120th meridian approaching the mountains. e011205457, Library and Archives Canada

pography. I shall proceed to complete the survey north of the Wapiti River as soon as possible.

The crew moved across the Narraway the next day. As he had done around the Wapiti, Cautley spent a couple of days surveying the surrounding topography and measuring elevations. South of the Narraway, the 120th meridian line rises to 2097 metres as it crosses a mountain Cautley named Torrens. Cautley established station 61-4 at the highest point on Torrens along the 120th meridian. He also set a station and built a large cairn at the summit of Mount Torrens. In addition, he established two stations, A and D, that could be used as part of a triangulation survey. Cautley described his surveying in a letter to Deville.

> The line has been left on a high outline ridge of a big mountain which occupies an area of perhaps 25 square miles and lies between encircling branches of the Crooked River. I have called the above mountain Torrens Mountain after the Australian lawyer and politician who introduced the Torrens system of Land Titles in 1858 and shall be glad if the above name can be submitted for the approval of the Geographic Board. Torrens Mountain is only 7400 feet high [2255 metres] or rather more than 4000 feet [1219 metres] above Crooked [Narraway] River, but it is higher than anything else in the immediate vicinity and higher also than many peaks to the south. Unfortunately dense smoke prevented us from taking photographs or making observations ahead. The end of the line is marked by an I.P. (61-4) and cairn near the middle of Tp. 61. There is also a big cairn on the summit of the mountain and two other triangulation cairns from which the work can be carried on.

The smoke became so thick that it interfered with the surveying. On July 27, Cautley wrote in his diary: "Went up to 62-1 where I read final azimuths

Conference held in October 1921, as contained in their report. An additional reason for stopping my survey at the above point is that the mountainous character of the surrounding country will necessitate the use of the photographic method surveying in order to get in the required to-on A & D and elevation, but was unable to get photographs or topographical information on account of dense smoke." The next day, "smoke so bad that it was impossible to do anything." On July 29, Cautley wrote: "Smoke as bad as ever; even the outline of mountain ridges 2 miles [3.2 kilometres] away invisible. Climbed with whole party to summit Torrens, selected site & collected rock for cairn & read some elevations on nearby points and worked in a lot of detail topography." The men had a Sunday in camp on July 30, while Cautley spent the day plotting topography. On the last day of the month, "self read azimuths, elevations and topographical intersections at Torrens summit and 61-4. Built cairn 27'5" [8.3 metres] around base & 8'0" [2.4 metres] high, best I ever saw. Mr. Gorman read azimuths at 61-4 & 61-5 & built good cairn at 61-4."

On the first of August, a bush fire started on the southwest slope of Mount Torrens, so the camp was moved back to the north side of the Narraway River. Before finishing their surveying in the area, Cautley and Gorman needed to go to stations A and D, which they had not finished on July 27. On August 2, Cautley wrote: "Smoke very dense and Mr. Gorman and self had just arranged that he would stay behind with one man and finish up at D and A when about noon it suddenly cleared amazingly (considering that wind still remained in southwest). Mr. Gorman and self got horses at once; he went to D and I to A and we both managed to get through and returned after dark." Cautley also noted: "Torrens fires so bad that we could not have remained on Torrens today."

On August 3, the Alberta crew began heading north, and arrived on the north bank of the Wapiti on August 5. "Smoke bad and smell of fire. Wapiti is 3.15 feet [1 metre] lower than when we crossed in June and ford is 1.66 feet [0.5 metre] deep in shallowest water."

Since the survey line was already clear for surveying the 120th meridian north, Cautley did not need his full crew for the second part of the summer, so five men were paid off. On August 7, Gorman and the crew started surveying north, while Cautley and one of the packers took the five discharged men into Rio Grande and arranged for their transportation to Grande Prairie and Edmonton. Cautley purchased some supplies in Rio Grande.

Crossing the Narraway River. e011205458, Library and Archives Canada

By August 12, Cautley was back in camp and pleased with the work done in his absence. The surveying proceeded smoothly through August, and many of Cautley's diary entries state: "Levelling, chaining, line clearing, monument building and packing all day."

Rio Grande post office (left) and store. Ewen MacDonell was the nephew of the builder of the ED & BC Railway, which had a station in Grande Prairie. In 1921 he and Alex McIntosh opened the first store in Rio Grande. The men brought their supplies from the rail line in Grande Prairie. e011205461, Library and Archives Canada

Cautley observed: "Owing to extreme dryness of season it is possible to work horses over muskeg which was altogether too soft in 1920; from same cause the trail is rotten with nests of ground hornets, which upset the horses."

On September 2, Cautley wrote: "Since August 21, 22 monuments have been built in 12 days, involving over 60,000 pounds [27,215 kilograms] monument material packed on horses." In his government report, he noted that "every monument built on the 120th meridian is of full standard size and requires 3000 lbs [1360 kilograms] of gravel, cement and water for its construction."

By September 9, the work was finished. Cautley wrote:

Men have done splendidly. Bill, Ross and Johnny have done the packing and distributing. Mark, Fred and packer Henry have built two complete monuments every day—much more difficult than it sounds. Mr. Gorman and John have done the leveling, Gerald and Nicholson chaining. Ed and Bob cut out the line, dug walls for the monuments and put up camps, while I have acted as rover.

On September 12, the crew was back in Rio Grande, where Cautley arranged transportation to Grande Prairie.

Cautley's work was not finished yet. Since he was going to survey the 120th meridian north in 1923, he

Transportation in Peace River country. e011205462, Library and Archives Canada

needed to winter his horses in the Peace River region. With Johnny Napoleon and two other men, Cautley headed for Fort St. John. He also hired a democrat with a team of horses. On September 14, "got packed up and ready by noon, including getting another democrat, the one provided not being strong enough in my opinion. Made 16 miles [26 kilometres] with team, democrat & 32 loose horses." The next day, the men reached Swan Lake. While there, Cautley relevelled a section of the boundary line "and discovered error which has been under discussion for three years." By September 19, the men had reached the crossing of the South Pine River. "Ferry is working. Ferried saddlery across in small boat, swam horses, resettled and delivered horses to Henry W. Philip, 2 miles [3.2 kilometres] back from river, just at dusk."

Cautley started the return trip the next day and reached Pouce Coupe, where he paid off Johnny Napoleon. On September 21, he arrived at Ray Lake. "Policeman watching the night road for suspected bootleggers made the tent a kind of rallying point. I hoped there would be no promiscuous shooting, and there wasn't." On September 24, the men arrived in Rio Grande. The next day a man and truck arrived from Grande Prairie. "Self, two men and about 1600 pounds [725 kilograms] of freight arrived in Grande Prairie after various breakdowns." Cautley spent a few days in Grande Prairie sorting his supplies, storing items he would need next year and sending four cases of equipment to Ottawa. On September 29, he took the train to Edmonton, finishing his 1922 field season.

Cautley's photo of "gas boat *Weenusk* leaving Dunvegan." C-026392, Library and Archives Canada

1923

Two of the main people who had been involved with the boundary survey since its beginning in 1913 had a diminished role in 1923. Each year since 1920, A.O. Wheeler had spent less time doing the actual surveying, while A.J. Campbell took increasing responsibility for the fieldwork. In 1923, Wheeler (as he had also done in 1921) hired an assistant to help Campbell with the surveying. He selected A.S. "Spike" Thomson, who had been on the BC crew for four years and was familiar with the project. (Unfortunately, there is no existing diary for Thomson as there had been for the previous years.) Twice in 1923, Wheeler joined the crew but both times unforeseen circumstances limited his time in the field.

In 1923, Canada's surveyor general, Édouard-Gaston Deville, was in his seventies and in declining health. Although Deville retained overall supervision and involvement in the projects handled by his office, his assistants handled much of the daily work. During this year, Deville assisted Cautley's survey by providing a person to run the levels, along with an assistant to the leveller. J.N. Wallace, federal commissioner for the boundary survey from 1913 to 1915 and director of levels, had overall supervision for this aspect of Cautley's work. Deville also continued the Geodetic Survey's support for the BC survey along the Great Divide.

There was a winter project again in 1923, this time involving the Geodetic Survey. And there was personal tragedy in Wheeler's life.

WHEELER AND THE BC SURVEY CREW

The Boundary Commission report described the beginning of the BC crew's field season:

"The survey had been discontinued the previous season at Bess Pass, distant six days by pack train from Jasper, and as the party would be constantly getting farther away, it was necessary to get supplies forward to the field of operations as soon as possible." The crew organized in Jasper from June 7 to 9.

On June 10, the assistant, Mr. A.J. Campbell, DLS, moved out with eighteen head of pack horses loaded with supplies to establish a cache at Colonel Pass, situated about halfway. The streams were in heavy flood and one horse drowned and a load of supplies was lost when crossing Meadow Creek. In addition to this bad piece of luck, eight head of horses wandered away looking for food, of which there was very little early in the season. The three days lost from this compelled Mr. Campbell to cache his loads when two days out, and he returned to Jasper on the fifteenth. Mr. Wheeler arrived with the full party the same day and another pack train was got ready. A start was made June 18, but again delay was caused by missing horses wandering in search of food.

Wheeler had originally intended to join the crew at the start of the surveying, but with the delays, there would not be sufficient time before the start of the Alpine Club camp. He returned to Banff, where he wrote to Umbach on June 20: "Last summer we were troubled by bush fire smoke. This year we are troubled by floods. When taking out supplies a few days ago, Mr. Campbell had a horse drowned and a load of supplies lost.

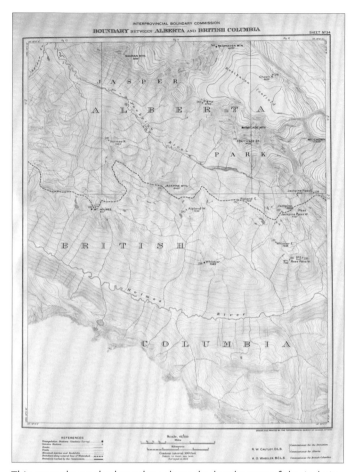

This map shows the boundary along the headwaters of the Jackpine River. "Boundary between Alberta and British Columbia" map atlas, Sheet 34, Land Title and Survey Authority of British Columbia

The water is falling now and the snow going very rapidly which is good."

On July 6, almost four weeks after originally departing from Jasper, Campbell and the crew arrived at the mouth of Carcajou Creek, the area near Bess Pass that the BC crew had been unable to survey the previous September. The following day the crew travelled to Carcajou Pass and from July 7 to 12 surveyed five stations in the vicinity. Campbell finally reached Bess Pass on July 16 and surveyed two additional stations in the area the following day.

Now the BC crew was ready to resume surveying towards the 120th meridian. They crossed Jackpine Pass and during the rest of July occupied fourteen stations "on the heights bordering the Jackpine River val-

ley and westward towards the Holmes River valley, in close proximity to the main watershed," the Boundary Commission report stated. It was the most productive part of the 1923 field season for the BC crew. The crew surveyed from three of Lambart's Geodetic Survey stations, establishing a good connection between the two surveys.

This was followed by rain and snow for most of the first ten days of August, when only four stations were surveyed. Then time was spent cutting trail, and on August 14, the camp was moved to Loren Lake pass. From August 15 to 20, eight stations were surveyed in the area.

Meanwhile, Wheeler came to join the crew. On August 8, he arrived in Jasper where packer Loren Nejedley met him. They departed around noon the next day, going up to Yellowhead Pass and then following the route along the divide that Campbell and his crew had

Part of the panorama of the Resthaven Icefield. v771_pd_49_168_001 and v771_pd_49_169_001, Whyte Museum of the Canadian Rockies

cleared. By August 13, they had crossed Moose Pass, where they met one of Lambart's Geodetic Survey crews; reached the ford across Calumet Creek, where they met Fred Brewster; and moved on to the Mount Robson flats.

Wheeler spent two days there. He wanted to make a preliminary examination of the divide through the Robson Pass area, because Cautley was planning to survey it next summer. In addition, Wheeler wanted to look over "possible campgrounds for next year's ACC." He was already thinking about having a special ceremony next summer at Robson Pass to celebrate the completion of this part of the Alberta-BC boundary survey and to combine it with the 1924 Alpine Club summer camp at the same location.

On August 16, the two men resumed their journey. They reached Bess Pass early the following morning, crossed Jackpine Pass and stopped at the same location where Wheeler had camped with Lambart the previous year. The men travelled down the Jackpine valley on August 18. "Very bad going," Wheeler wrote. "Stopped at 3:30 P.M. Loren Nejedley got kicked by a horse." The next morning they found a note from Art Hughes and at noon a note from Campbell with directions. Wheeler and Nejedley stopped to camp at 5:00 pm without reaching Campbell's camp. After less than an hour's travel the next day, they arrived at the camp. Campbell was out on a fly camp, so some men went to get him, and Campbell returned later in the day.

The crew moved camp to a tributary of the west branch of the Jackpine River on August 21 and remained at this place the next day. "During afternoon had visit from Mr. Kilpatrick of Lambart's party," Wheeler wrote in his diary. "Camped about a mile or more from us. Lambart left us some sheep which we went and got. Gave Kilpatrick some tobacco. They were out." Lambart and his crew had finished establishing the Geodetic Survey stations and were on their return trip. On August 23, the crew moved to Beaverdam Pass.

During the last week of the month, low clouds and rain limited the work to five stations. On August 25, Campbell and a group of men surveyed from one of Lambart's stations, while Thomson and one man went to one of their own stations. Wheeler and Nejedley went to Avalanche Pass and set a signal in the morning; in the afternoon, they "sorted out stuff to leave at cache on west branch Jackpine crossing for trip home."

Wheeler joined Thomson and a group of men on a trip west of their camp on August 27. In his diary, Wheeler wrote that they had just found a suitable campsite when a crew member came from camp "with a letter from Mrs. Brewster with copy of telegram that Clare [his wife] was very ill [and] to come at once. Return to camp immediately. Got ready to go to Jasper in the morning." Wheeler camped the first night along the Smoky River. The shortest route to the railway was to follow the Mount Robson trail:

> Made Robson camp for lunch. No word to be had there. Started for Robson Station. Soon after met packer who told me Clare was dead. Went to Robson Station and took train to Jasper. Mr. and Mrs. Brewster

Top: North of Kicking Horse Pass the only section of the boundary that the Mountain Legacy Project has rephotographed lies north and west of the Resthaven Icefield. The 1923 photograph from this station west of Resthaven Mountain shows the glaciers and snowfields of the area at that time. WHE1923_W23-206, Library and Archives Canada

Bottom: The Mountain Legacy photograph in 2007 shows a small decrease in the extent of glaciation. Vegetation extends farther into the basin in the front centre of the picture. MLP2007_W23-206, Mountain Legacy Project

Left: This view from Draco station looks over the Jackpine River valley. WHE1923_W23-223, Library and Archives Canada. Right: In many of the repeat photographs along this section of the boundary, the changes that have occurred in the landscape in the past century are more gradual. Landscape changes in the river valley are visible. MLP2007_W23-223, Mountain Legacy Project

met me and took me to their home where I got letters and telephone. Spent the night at Jasper sorting out baggage and getting business done. Got a few hours sleep.

The next morning he departed for Banff.

At the beginning of September, rain and snow limited surveying. From September 6 to 8, Campbell's crew surveyed four stations, two of them established by Lambart. On September 10, the BC crew moved to Fetherstonhaugh Pass and surveyed nine stations by September 14, one of them a geodetic. The surveying season finished at Casket Pass.

Campbell's crew started their journey back to Jasper on September 16. They returned via the Mount Robson route and continued by pack train up the Fraser River valley, stopping for a day to survey two stations on the west side of the Moose River valley near its junction with the Fraser. The crew reached Jasper on September 28. Despite the difficulties at the beginning of the season, the bad weather in early August and Wheeler's personal tragedy, it had been a successful season, augmented by the absence of smoke, which made the days clear for surveying. The BC crew surveyed sixty-one stations, made many connections with the Geodetic Survey's network and were within close enough distance of the 120th meridian to complete their work the next year.

Tree tablet at a camp near the mouth of the Jackpine River. PA64-2-A, Jasper Yellowhead Museum & Archives

H.F. LAMBART

The Geodetic Survey had a winter project in 1923. On January 12, surveyor K.H. Robb started establishing the southern baseline for Lambart's network. In the Geodetic Survey's annual report, he wrote: "The rough nature of the country made it necessary that the base line be run down the centre of Yellowhead Lake and that the work be done when the lake was frozen over." A concrete pier was built over the geodetic station that Lambart had established at the east end of the lake. From this station, a straight line was measured to the concrete pier at the west end. Posts were set in the ice at 50-metre intervals. First, the crew cut holes in the ice 25 centimetres deep. Then they put the posts in position, packed snow and water around them and let them freeze overnight. There was open water at the narrows near the west end, so a deflection of 6 degrees was necessary for the last part of the baseline. Once the distance was measured, levels were run.

The measurements could be made only when the temperature was above -19°C, so there were days when the crew could not work. The coldest temperature was -36°C. Robb completed the baseline on February 10.

Lambart had three crews working on the Geodetic Survey network during the summer of 1923. There were two crews to measure the angles from the survey stations. Since the crews were only reading angles, both consisted of just a couple of men and a packer. Lambart was in charge of a crew that would finish establishing the stations for the network. Felix Plante was his packer and guide again. Only Lambart's diary for the first month of this field season remains.

Lambart spent most of June in the Yellowhead Pass area at the southern end of his triangulation network. He arrived in Jasper on June 8 and got all three crews organized. On June 14, one of the survey crews went to Yellowhead Pass, while Lambart accompanied the other to Robson Station. The men spent the next day moving supplies for both crews, and Lambart returned to Yellowhead Pass in the evening. At the pass, Lambart put a signal over Cautley's initial 1917 monument in Yellowhead Pass. He spent a couple of days clearing line, surveying from the piers established at Yellowhead Lake, and "occupied the point where the base makes a deflection." A couple of men climbed Barry, one of their stations south of the pass, and placed a signal at this location. Lambart went to Tête Roche, their first station north of the pass, where he found that he needed to add another station for a better view to the north. On June 22, Lambart and his men travelled south to Mount Fitzwilliam and stayed at the campground Wheeler had previously established along the trail. It rained on June 23, but the next day Lambart and

Left: One of the photographs at the Jones Pass West station. WHE1923_W23-255, Library and Archives Canada. Right: The 2007 repeat photograph shows some changes in the vegetation. MLP2007_W23-255, Mountain Legacy Project

the men got an early start and spent six hours on the summit. Mount Fitzwilliam is the highest peak around Yellowhead Pass, and the BC crew had surveyed there in 1917, so it provided an important connection for Lambart.

After completing the surveying around Yellowhead Pass, one of the crews began surveying the geodetic net-

work from that location. On June 26, Lambart's crew and the second survey crew returned to Robson Station, where the packers were waiting. They departed the next day, and, by June 29, were at Calumet Creek at the "old campground with the bath and very comfortably fixed. In one of the heavenly spots of the mountains, place a blaze of bloom." During the last day of June, Lambart and three men climbed Calumet Peak.

Plante packing a horse. e011205427, Library and Archives Canada

When he surveyed from the station, Lambart found that his transit wasn't working properly. July 1 was a Sunday. "Glorious soak in the pool and a sound sleep." The next day, the second survey crew continued to their starting point, while Lambart's crew moved back to Robson Pass. Lambart wrote: "The instrument has been thoroughly overhauled and adjusted, and if we don't get closures we will have to abandon the instrument and delay the work until McGrath can get mine back."

One of the people on Lambart's crew was Rev. George Kilpatrick, a minister from Ottawa whom Lambart knew personally. He had no surveying experience, so he helped build cairns and assisted wherever possible. Kilpatrick joined Lambart's crew at Robson Pass in early July. Twice in the first three days they climbed Mumm Peak, one of the main stations in the area. On July 10, Lambart and his crew travelled north to the place where he had stopped surveying in 1922. They began establishing new stations on July 15. From that date until August 10, the crew established eleven stations to complete the geodetic network. In addition, they surveyed a baseline at Jarvis Pass that they initially believed would be near the intersection of the 120th meridian.

However, during the surveying, Lambart made an important discovery. From Cautley's 1922 surveying calculations, he knew approximately where Mount Torrens would be, and he was able to locate the large cairn on the isolated peak. In the 1923 Geodetic Survey annual report, Lambart wrote that Cautley's cairn on Mount Torrens was "located with reference to our triangulation, and the intersection of the meridian and the Continental Divide was thereby found to be near what is known locally as the Sheep Creek pass, 18 to 20 miles [29 to 32 kilometres] south of Jarvis Pass." This was the Geodetic Survey's first connection to the 120th meridian. Since the intersection of the Continental Divide and the meridian was much farther south than it had been thought to be, it would necessitate changes to the connections made by the three different surveys in 1924.

R.W. CAUTLEY

In 1923, Cautley spent one more season surveying north along the 120th meridian, beginning where his 1919 surveying had finished, north of the Peace River. His objective was to survey as far as possible. To expedite the work, surveyor George Palsen and an assistant ran the levels. All of the stations were to be marked with dirt mounds instead of concrete monuments. This was a fortuitous decision, since only one poor source of gravel was located during the season.

This section of the 120th meridian had only three notable geographical features: the Clear Hills, the Doig River and the Chinchaga River. The Boundary Commission report stated:

> The dominating characteristic of the whole country traversed in 1923 is muskeg. Comparatively high ridges are generally visible on the skyline and form local watersheds between the various drainage areas of the plateau, but these ridges are many miles apart and the intervening country is extraordinarily flat. So flat is it that although the surface is by no means level—and indeed varies considerably in elevation—the drainage is sluggish and indirect to such an extent that the land has become saturated with water and, in the course of many centuries, has developed a sphagnum moss character.

The report also noted: "Owing to the muskeg character of the country, some difficulty was experienced in connection with horse transport."

The easiest access to this section of the 120th meridian was by boat. Cautley made arrangements with the Hudson's Bay Company to have their motor boat, the *Weenusk*, transport the men and supplies from the town of Peace River to Boundary Landing, about 5 kilometres downstream from the 120th meridian.

Cautley, Gorman, Palsen and a crew of twelve men left Edmonton by train on May 17 and arrived at Peace River the next day. The *Weenusk* returned from a trip downstream that night. The men spent May 19 loading

Glaciers near the summit of Mumm Peak. e011205426, Library and Archives Canada

the cargo, and the boat departed in the late afternoon. It was pushing a barge that was 23 metres long and 4 metres wide and held 30 tons of supplies. Two days later, the *Weenusk* arrived at Boundary Landing, where the packers were waiting with thirty-one horses.

On May 22, the crew moved camp 10 kilometres, and by May 23, after a move of 20 kilometres, they were camped at the north end of Boundary Lake, where their survey of the 120th meridian would begin. First, Gorman and Palsen took the crew and spent a few days

George Palsen, leveller. Palsen was born in Iceland and came to Canada when he was young. He became a certified Dominion leveller in 1916, so he had special qualifications for surveying elevations. Later he became a licensed surveyor for both the federal government and in Alberta. Most of his career was spent on surveys in northern Canada. Palsen also worked on the final section of the boundary survey in the 1950s. e011205464, Library and Archives Canada

making a special survey at Boundary Lake requested by the federal government for a proposed drainage plan. Cautley and one man surveyed a traverse of the lake, while the packers brought the supplies up from the river. In his diary, Cautley noted dead rabbits everywhere from one of the periodic die-offs of this animal.

On May 28, after a Sunday in camp, Cautley finished the traverse of Boundary Lake, while two men began making the mounds for the special survey, and Gorman and five men started cutting line ahead. By the last day of May, all the supplies had been brought

from the river, the mounds for the special survey were finished, and the trail had been cleared a sufficient distance to enable the crew to move to a new camp 10 kilometres ahead.

The next day, they "struck big marshy lake at 4 PM and planned base for triangulation." It took a couple of days to survey the line and make a triangulation of the lake. On June 5, the crew moved camp about 7 kilometres north along the line. "Good water but no horse feed." Throughout the field season, it was difficult to find locations with good feed for the horses. At this

The summit of the Clear Hills. One of the dirt mounds used to mark the stations in 1923 is visible. Behind the man is a signal because the station could be seen from a long distance away. e011205463, Library and Archives Canada

camp, Cautley and Gorman made their first observations on the stars and found that their line was about 20 seconds in error. On June 8, the two men made observations once more, but again, the results were not satisfactory. (Cautley does not provide any explanation for the discrepancies, but the soft ground may have made it difficult to keep the transit stable.)

Through June, the work proceeded. In the middle of the month, the crew encountered a large area of muskeg along the line. They were able to survey the line, but it took time to find a trail to get the horses through. A few days later, the crew encountered another section of muskeg. On June 19, Cautley wrote that he had located a new camp that was 11 kilometres along the line but 21 by trail.

On June 21, the 120th meridian reached the summit of the Clear Hills, the highest point on their survey in 1923, with a view for many kilometres. The next day, one of the axemen cut his foot and Cautley had to "sew it up." While making observations on June 23, two bears came within 20 metres of Cautley. He made observations on June 26, and the results seemed to be satisfactory. Cautley had made arrangements to pay a settler along the Peace River named Streeper to bring the mail in once a month. He arrived on June 29 and stayed for a couple of days. On Sunday, July 1, Cautley wrote to Deville:

> This is the first opportunity I have had of sending mail outside since we started. The party reached Boundary Lake on the 23rd May. The first work done was that in connection with the special survey at Boundary Lake referred to in your instructions....
>
> The special survey was made by Mr. Gorman, DLS, who will report on it and put in final returns when he gets back to Ottawa. The above survey occupied three full days of the whole party's time....
>
> Work on the boundary survey proper commenced on 28 May, since which time 27 miles [43 kilometres] of the survey have been completed. Just at present both line and trail are in very heavy going, but I hope and expect to get into easier country shortly.
>
> With the exception of one man who cut his foot badly 10 days ago, the whole party is well.... The horses are doing well, although owing to absence of feed, the horse camp is at present 8½ miles [13 kilometres] south.

There was rain during twenty days in July, and several days of work were lost. On July 3, it rained in the morning but stopped by noon. Cautley wanted to work in the afternoon, but several men refused, saying it was too wet. Cautley threatened to deduct a day's wages for those who stayed in camp. In his diary, Cautley wrote that the next day

> nine men came in and demanded their time unless I would allow them yesterday's wages. This would have absolutely ruined the season's work, as it would have been most difficult to get new men at the beginning of the hunting season and would cause loss of at least two weeks.... In the interest of the work therefore I told

them that if they would undertake that Mr. Gorman and myself should be the sole judges as to whether weather conditions justified work in future I would accede to their demand as to payment for yesterday.

The crew members agreed, and there were no further difficulties during the field season.

For July 6, Cautley wrote: "Mr. Gorman cut his foot, fortunately not very badly. Jack Idle, a local trapper, was in camp." The next day, the crew "struck some good going," but by July 9, with the temperature at 32°C in the shade, they were back in the muskeg all day. This was an especially difficult section, and the men had to pack the supplies they would need for the next few days while the horses went "round by a trail which runs ENE [east northeast] to the Doig River and then switches west on the trail which is said to follow the Doig River."

The temperature was 32°C again the next day. "Very hot all night and flies worst yet; nobody slept and it looks like a big storm—constant thunder but no rain. The great difficulty is producing, even with the tripod on hubs 5 feet [1.5 metres] in the muskeg. Ordinary production is out of the question, besides which natural light is only good after 8 PM." The pack train left on July 11 with four axemen to clear trail. In the evening, Cautley went back about 5 kilometres "and succeeded in producing line by lining in on back stations and 88-3 (summit of Clear Hills)."

Gorman left on the morning of July 12 to try to locate the Doig River, "returning late in evening—a tremendous relief as we did not know with any certainty where the river was." Cautley wrote: "All will be well if only it doesn't rain. Nothing but level, wet muskeg everywhere." The next day, the crew moved camp about 3 kilometres "to an island of big poplar." The cook "managed to burn up almost completely his fine big second year tent by dumping ashes from the cook stove just outside the door and moving off for a minute and leaving it." The next day they packed into "Doig River and just caught pack outfit as they were preparing to move on having missed Mr. Gorman's signals. Rained hard all afternoon. Made 7 miles [11 kilometres] this week and pretty tough sledding. The fact that it did not rain until we rejoined the main camp saved the situation."

Cautley had trouble with bears getting into the food. On July 2, a "big black bear made an awful mess of the cache at camp 9. George and Harry went down in evening and shot him as he was approaching the cache but failed to kill him outright although badly hit." On July 18, he wrote: "It appears a bear managed to get on top of the tree cache in camp 10 and did more damage." The next day Cautley "made inventory of everything in the cache and find that I must send a man back for supplies, 450 pounds [205 kilograms], to remedy bear degradation."

Cautley sent a letter to Deville with a couple of packers who went to Fort St. John to pick up supplies.

I am camped on the Doig River, or as I am inclined to think, a more northerly main tributary of the North Pine, just north of the north boundary of Township 91. In townships 90 and 91 there is one stretch of line 10 miles [16 kilometres] in length which does not touch sound ground at any point. In order to get here we man packed 11 miles [18 kilometres] by line and sent the horses around, a distance by trail of about 24 miles [40 kilometres]. Immediately north of camp there are 5 more miles [8 kilometres] of muskeg, but I have found a trail around it and from the next camp—5¾ miles ahead—the country rises considerably so that I hope our muskeg difficulties are, in the main, over.

While camped along the Doig River, Cautley made a survey of the area. The Boundary Commission report describes the valley: "Doig River drains a vast muskeg country; the valley at the boundary is 75 feet [23 metres] deep and it is characteristic of the country that within twenty feet [6 metres] of the top of the banks the surface stretches away in wet muskegs for miles."

On July 23, "Mr. Gorman got three observations in evening which we need badly and none secured since big muskeg." The next day, the crew left the Doig River. "At 11 AM a thunderstorm accompanied by a deluge of rain began and lasted until 11 PM. Party was 4 miles

[6.4 kilometres] north of camp 12 and pack train just started. Both arrived in camp 13 at 4:55 PM amid a fresh burst of rain. Not a dry stitch in camp." The storm continued on July 25. Cautley wrote the following day:

Camp wet, beds wet, and all the clothes in camp ditto, so although it rained very little, I didn't go out at all today. I don't think I ever saw so much rain fall in 36 hours as on the two previous days and the whole country is running water. Made up topography sheets, worked out Mr. Gorman's three observations of the 23rd, which came out excellently.

The crew headed north towards the Chinchaga River on July 30, moving to their next camp. "Trail very wet and some horses down," Cautley wrote. The Boundary Commission report stated:

There are a number of Indian trails throughout the country, most of which seem to be of ancient origin.... Most fortunately for the success of the season's work, a tributary of Doig River, which Mr. Cautley named Adskwatim Creek [meaning many dams in Cree], closely parallels the Boundary for almost the entire distance of 26 miles [42 kilometres] between Doig and Chinchaga Rivers, and there is a fairly good Indian trail following it which served the purposes of the survey admirably. For miles on either side of the Boundary the surface between the two rivers is so largely composed of muskeg that it is doubtful if another trail could have been found close enough to the work to be of any service as far as the horses were concerned.

On August 1, Cautley "located camp No. 15 4¼ miles [6.8 kilometres] ahead on line and such that we can use the existing trail along the creek all the way. Struck three live beaver dams." The crew moved to this camp on August 4. Johnny Napoleon shot a moose. The next day, some men brought in the meat and the

9-1

S.6683

"Streeper's squatting claim on Peace River three miles [4.8 kilometres] from boundary," Cautley wrote. e011205465, Library and Archives Canada

packers returned from Fort St. John. On August 8, the crew moved about 5 kilometres north along the line.

The next week was spent chopping line, and on August 15, the Alberta crew moved 13.6 kilometres to the Chinchaga River. "All horses got across the muskeg without mishap but cut it up pretty badly." On August 17 and 18, Cautley surveyed the topography along the river.

The crew continued north on August 20. "Moved 6½ miles [10.4 kilometres] by trail and 3½ [5.6 kilometres] by line to camp No. 18. Bad trail; horses down and flour and rolled oats spoiled.... Self took some more topography along Chinchaga, measured stream, helped pack etc." The next day, one of the packers went ahead to locate the final camp. He found that there was a lot of muskeg, making it a difficult section to travel. On August 24, the crew moved camp. "The horses got through pretty well but cut up the trail badly in places." The next day the packers repaired the trail for the return trip.

From this camp, Cautley continued the line north for about 3 kilometres. "George Baker came within a fraction of being killed by a falling dry poplar but providentially fell between two logs on the ground and escaped with a blow on his arm." On August 28, the final station, 97-5, was built. It "occurs on the summit of a high ridge and overlooks toward the south nearly 50 miles [80 kilometres] of the low, flat country through which the survey passes."

The return trip started the next day. The men "noticed on the way back that a great many of the carefully built mounds at monuments had been partially destroyed and tumbled into the trenches" by bears. By August 31, the men were on the south side of Doig River. "Trail very wet and horses tired," Cautley wrote. On September 3, they camped on the south side of the Clear Hills. "Only horse feed on trail within 26 miles [43 kilometres]. Out of sugar this morning, having eaten 260 pounds [118 kilograms] in 29 days." Two days later, they were on the south side of Boundary Lake, where Cautley purchased 36 kilograms of sugar from a settler. On September 6, Cautley went ahead to Streeper's farm on the Peace River. He purchased a big calf from Streeper, who butchered it for him. The *Weenusk* had gone up the Peace River early in the morning, leaving mail for the crew.

The horses swam the Peace River on September 8, heading with the packers to spend the winter around Rio Grande. The *Weenusk* returned late in the evening of September 9 and tied up for the night. On the tenth, the boat left early and arrived at Peace River fourteen hours later. Cautley paid off the crew as of September 12, when they arrived in Edmonton. He packed his surveying equipment to go to the surveyor general's office in Ottawa. Then he travelled by train to Grande Prairie, arriving on the afternoon of September 13. "Found by telephone that packers had arrived with horses at Rio Grande and immediately went out there in a car. Checked over stuff with MacDonell, had supper with him, and returned to Grande Prairie with two packers whom I paid off and dropped at Lake Saskatoon

at their request." The next day, he took the train to Edmonton, arriving on September 15.

It was a successful season for Cautley. He surveyed about 125 kilometres of the 120th meridian, extending it past 57 degrees latitude. Seventy monuments were established; levels run for the entire line; topography taken for about 2 kilometres on each side of the line; and connections made with two baselines. This was the farthest north that the 120th meridian would be delineated for the foreseeable future. There was no immediate need to define the remaining portion of the boundary because there were very few people living in the area, few apparent natural resources, and most of the country was muskeg.

Cautley's horses brought in for oat rations at Miette Pass in September. e001205430, Library and Archives Canada

SURVEYING THE 120TH MERIDIAN AND THE GREAT DIVIDE

1924

The Alberta-BC boundary survey entered its twelfth year in 1924. The three levels of government, and the people who had been involved with the project since its inception, were looking forward to the conclusion of this monumental survey. In January 1924, Wheeler, Cautley and representatives from the Alberta, BC and federal governments attended a conference in Edmonton. They needed to plan for the conclusion of the surveying, prepare for production of the final report and make formal arrangements for the Boundary Survey Commission.

The first and immediate priority was to finish the survey in 1924. Cautley had to delineate the 120th meridian line south until it reached the Great Divide, and Wheeler needed to survey along the Rocky Mountain watershed until he intersected Cautley's survey. The remaining time would be used by Cautley to survey some passes, while Wheeler's crew would attempt to complete the phototopographic survey to Mount Torrens. During the fall of 1924 and into 1925, Wheeler and Cautley would complete their calculations, produce maps, label photographs and write their final reports.

Before the Alberta-BC boundary survey commenced, all three levels of government had to officially approve the project. The same procedure was needed to conclude the survey. The conference report stated that

> on the completion of the publication of the report and atlases, recommendation should be made to the executive Councils of the provinces of Alberta and British Columbia that the reports of the commission be approved and the commission be disbanded. On the passing of the necessary orders by the provincial governments, the Dominion Government shall provide by

an order of the Privy Council or by legislation if considered necessary, the ratification of the report and the definition of the boundary on the ground.

After 1924, the northern part of the Alberta-BC boundary remained to be surveyed sometime in the future, and it was likely that there would be ongoing issues related to the demarcation of the line. The conference report also urged that

> in order to deal with all questions in connection with replacing of all monuments which may be destroyed or disturbed, the re-adjustment of any portions of the boundary and the dealing with any matters concerning the boundary, it is recommended that the surveyor general of Canada, the director of surveys of Alberta and the surveyor general of British Columbia be constituted the standing committee to which all such matters may be referred.

As in the previous two years, there was a small special project before the beginning of the regular surveying, and a new technology—one that would have an important influence on surveying—played a small but vital role in the 1924 work of the boundary survey.

WHEELER AND THE BC SURVEY CREW

Once again, A.J. Campbell had a surveying assistant—N.C. Stewart, BCLS #120. Wheeler's initial focus for the 1924 field season was the Alpine Club summer camp at Mount Robson during the second half of July, and the ceremony for the boundary survey that would

Casket Mountain (originally called Coffin), named for its unique rock formation. WHE1924_8-A-O-W-1924-E, Mountain Legacy Project

There is almost no snow in the repeat photograph, but this picture was taken later in the summer. The Mountain Legacy crew observed that several rocks on the casket had fallen off since 1924. They were also able to identify some ecological changes in the landscape. MLP2011_8-A-O-W-1924-E, Mountain Legacy Project

occur at Robson Pass during that time. As part of its tourist promotion, the Canadian National Railway agreed to pay for the cost of the monument commemorating the survey. Wheeler invited many dignitaries to attend the event. Initially, there was hope that Canada's surveyor general, Édouard Deville, might be able to be involved, but in the early spring his health declined precipitously, and by summer, F.H. Peters was appointed to this position.

Campbell and the BC party organized at Jasper and departed on June 23. They travelled via the Snake Indian River trail to the Smoky River, then up to Bess Creek. From here they followed the route used during the two previous seasons to travel along this section of the Rockies. On July 8, the BC crew arrived at Casket Pass, where they had finished working in 1923. During the remainder of the month, the crew occupied twenty stations.

On July 27, Campbell left to attend the ceremony at Robson Pass, and he did not return to his crew until August 14, leaving Stewart in charge during that time. Stormy weather limited Stewart's surveying to eight stations. During the second half of August, Campbell surveyed twenty stations. He was able to do this because the stations had already been established by Cautley and Lambart, and most were easily accessible. During that time, the BC crew completed the phototopographic surveying to Mount Torrens and surveyed at Intersection Mountain, where the Great Divide meets the 120th meridian.

Wheeler was occupied with several activities during the first half of August, but he wanted to be with the BC crew when they finished their surveying. He also needed to connect with Cautley to determine the watershed through Robson and Miette passes, the only two passes to be surveyed along this section of the Great Divide. They had been selected at the conference in January. The conference report cited two reasons:

> That both the above passes are in a part of the mountains that is now being used for tourist travel and will be much more extensively traveled in the near future, and on that account are more important than those which are more remote.

SURVEYING THE 120TH MERIDIAN AND THE GREAT DIVIDE

Cautley (left), Wheeler (centre) and Lambart at the special monument in Robson Pass. e011166448, Library and Archives Canada

That both the above passes are nearer to the railway than any others, and therefore that it will be more practicable to transport monument material to them in the very limited time that will be available.

The report discussed an inherent and apparent feature of Robson Pass:

At the present time there is no question of the definite location of the watershed of Robson Pass, as all the water discharging from Robson Glacier flows to Berg Lake in British Columbia, and cannot flow otherwise; but should the glacier again advance sufficiently the water from it will flow both to Berg Lake and to Lake Adolphus in Alberta, thus making the location of the watershed a very dubious operation.

On August 20, Wheeler departed Robson Station and arrived at Robson Pass. Cautley and his crew arrived on August 25, and the two men began examining the pass. The inspection continued the next day.

A view from the Sheep Creek South station. WHE1924_31-A-O-W-1924-X, Library and Archives Canada

Several changes in vegetation can be observed in the repeat photograph. MLP2011_31-A-O-W-1924-X, Mountain Legacy Project

Wheeler also visited with Lambart, who had completed the geodetic network and was on his way out. On August 27, Wheeler climbed to the upper slopes of Robson Pass and selected one of the stations there. He and Cautley established a station on the upper east side of the pass the next day. Snow kept the men in camp on August 29, and the two men spent much of the day visiting. On August 30, Wheeler and one man "climbed to 4-U and put in bolt below summit of Mount Mumm." Wheeler intended to leave to join the BC crew the next day, but his departure was delayed because two horses were not located until 5:00 pm.

Wheeler and packer Frank Finley departed on September 1. They camped at Bess Pass that night, remained there the next day because of rain and, on September 3, reached the Jackpine River. While travelling along the river the next day, "horse with my outfit took a swim. Took nearly 2 hours trying to dry out cigars and candies," treats that Wheeler was taking to the crew. On September 5, he crossed Big Shale Pass and reached McDonald's camp: "Very interesting travelling new ground for me." Wheeler and his packer arrived at a cache at Sheep Creek late the next afternoon. There he found a note Campbell had left two hours previously, so the two men continued on to the survey camp, arriving at 6:00 pm. "Boys put my tent up. Had supper and chatted a bit and turned in…. Distributed cigars and candies."

Around the Sheep Creek valley, the BC crew had some final work to do. Wheeler needed to connect the BC survey with the baseline established in that valley during the summer and also to survey at one of the nearby geodetic network stations. After a day of clouds, Wheeler attempted to survey at the geodetic station Haig: "I started to climb to shoulder of Haig. Soon commenced to rain so got under a tree and waited for it to clear. Waited until near 12 PM, so had my lunch and as it did not stop raining I returned to camp. Found Campbell and Reed had returned. Stewart's party came in later." It snowed and rained on September 9, but the next afternoon "went to the west end of the baseline and Campbell and Stewart read angles at end of base and at intersection of same with 120th meridian. Cold wind blowing. Clouds still down over higher places. Got most of the readings." On Septem-

One of the views from the final station of the BC photographic survey. v771_pd_66_491_001, Whyte Museum of the Canadian Rockies

ber 11, Wheeler attempted to reach Haig Ridge: "About 12 PM clouds lifted off the peaks and the parties were able to do their work. Stewart took station on south side Sheep Pass, Campbell one on north side. I reached the crest of the ridge but was cut off by break of rotten rock from Haig No. 1 station."

On September 12, both Campbell and Stewart occupied a station, "thus closing the work at Sheep Pass summit." Meanwhile, Wheeler "went out to the rim of the Northwest Basin and got a good view point a short way from Wallbridge. It gave me a good look out to the north. Tired. Turned in early. Glorious day, bright and warm." Wheeler probably had mixed feelings, knowing that the surveying had almost been successfully

completed, while realizing that this was his final field season and the end of his surveying career was near. The next day, the crew made a short move close to the east end of the baseline, while Stewart and Campbell did two final stations in the area.

On September 14, the phototopographical party started the return trip, heading for Miette Pass. Four days later, they reached the Calumet Creek area, where they stopped for a day, while Campbell went to Robson Pass for items that Wheeler had left there. Stewart, who was on his way out via Robson Station, accompanied Campbell. The return trip resumed on September 20, retracing the route Campbell had used at the beginning of the 1922 and 1923 field seasons. The following evening, the crew camped on the east side of Colonel Pass. A snowstorm started the next morning that lasted for three days. On September 25, the crew pushed on to Miette Pass. In his diary, Wheeler wrote: "Campbell, cook and I went ahead to break trail. Snow 3 feet [0.9 metres] deep in places, about a foot and a half [0.45 metres] on level. Deeper at summit of Grant Pass. The horse Colonel played out and fell down a slope. Tom left him behind. Camped beside Cautley. He and Mr. Gorman and his man helped us pitch camp."

Wheeler spent a day going over the surveying Cautley had already done at Miette Pass, and placed a bolt at the top of the pass. He had a toothache and probably wanted to leave the snow, for he departed at noon the next day, reaching Yellowhead Pass by dark. On September 28, the BC survey crew arrived in Jasper.

H.F. LAMBART

Lambart's objective for the field season was to complete the Geodetic Survey's triangulation network. He received permission to bring along his teenaged son, Arthur. Lambart had overall supervision, along with three small crews, each surveying a section of the network. One crew, headed by Don Fraser, worked on the southern section, from Yellowhead Pass to Moose Pass, and Lambart did not have contact with them during the summer. Hans Kihl and his brother John were each in charge of a crew that worked farther north.

On May 23, Lambart and Arthur arrived in Calgary, and left on the evening train to Banff with Wheeler. After a leisurely Victoria Day, Lambart spent the morning of May 25 "in conference with Mr. Wheeler, Logan [the 1925 Mount Logan expedition which included Lambart] and the work for the summer our only subjects." In the afternoon, Arthur had his first horse ride and fell off the horse. Lambart spent May 28 in Calgary hiring cooks and the following day in Edmonton, where he organized for the summer and visited Mr. and Mrs. Cautley.

On the last day of May, Lambart and his son took the train to Jasper. In his diary entry for June 2, Lambart wrote: "The [pack] train which left with provisions up the Snake Indian 15 May for the Sheep Creek pass has not yet been heard of. It was under the direction of Art Hughes with two others." Hughes was delivering supplies to be used when the crews worked farther north during the summer. Lambart also wrote: "Fred Brewster finds himself pretty shy of horses there being a number out on the ranges that he cannot find. Fred Brewster and I went down to the Mount Robson station and arranged with Dennison and Brittain for the use of his corrals and campground and teams to meet the freight on the 11th, when we were scheduled to arrive with the full outfit."

On June 11, "everyone and everything to Robson Station." Burt Kitchen was in charge of one pack train, Stan Kitchen another. Felix Plante was Lambart's packer. During the following days, the three pack trains moved all of the equipment and supplies to the Robson Pass area. Lambart wrote: "To my utter amazement the affair went off without a hitch except for the sad mishap to one of the packed ponies who ran a snag into itself, puncturing the intestines. The poor beast should have been shot at once, but instead we left it until the morning when we found it in terrible agony."

The packers constructed a cache on June 16, while two parties climbed and flagged Mumm and Lynx, two stations near Robson Pass. Lambart and surveyor John Kihl went to Lynx and "were gratified in seeing the flags gaily waving on Tub. We also picked up the cairns on Calumet and Plateau." The next morning Lambart received good news: "We were overjoyed in learning the presence of Art Hughes and his two men in camp a little way down. This gave us the opportunity of replacing the killed horse and swapping off the undesirables. A complete change consisted in our taking out six of Hughes and returning him five." Since Hughes was returning to Jasper to meet the BC crew, he could take the horses that Lambart did not want.

Lambart's crew and the two survey parties moved on June 17. It rained all day, "and we arrived in our campground early in the afternoon at the mouth of Carcajou Creek drenched." Inclement weather kept the crews in camp for the next three days. During that time, Arthur cut his foot. Lambart intended to continue travelling on June 22. John Kihl's crew left, but two horses could not be located. In the late afternoon, Lambart decided to take his crew to Bess Pass, leaving the other survey crew at camp.

The following day, Lambart encountered snow.

When Cassels and I arrived on the summit of Bess shoulder we found snow everywhere, much worse than any point south. The Jackpine Summit was full of it and [in] all the country stretching to the north between the Jackpine and the Beaver there was very little rock showing at all.… On the Jackpine we find the snow down to the very streambed and all through the timber. Felix and the rest got through all right and landed in the old campground about 4:30 PM. It's very delightful here this evening sitting at the open fire and watching the perfectly beautiful lighting on Bess.

The weather was fine during the next few days as Lambart and his crew travelled north. "We were favoured by another glorious day," he wrote on June 25. Hans Kihl and his party caught up to Lambart's crew that day. The two crews remained at the camp for another day, while Kihl and his crew surveyed a station. "Just a glorious day in camp getting properly cleaned and mended up and bathed. Hans with his two stalwarts, Tuite & Frame are making Resthaven, getting away on an early start.… Hans and his party did not get into camp until late having had a very hard time of it."

Lambart and Kihl's crews crossed Big Shale Pass on June 28.

On the summit Hans and I had a little sleep in the sun, on waking found Felix on the top with his train. The day was gorgeous and we had a superb unbroken view of the whole horizon. The study of the new point Resthaven was interesting. Hans' cairn could just be made out and appeared as if sticking out of the snow.

They camped that night at McDonald's campground where Lambart had stayed previously. The next morning, Lambart wrote:

I left with Arthur ahead of the train hoping to get a caribou and have a little companionship with the boy. We stopped to boil a billy… when the train overtook us. We parted with our good tea and drank water and in the end swallowed our lunch and chased after the trains which we stuck to the rest of the way. Everything seemed to be going well with the trains and Felix suggested that we go on to the Sheep Creek pass… to which I heartily agreed.

Lambart had arrived at the location where his work would begin.

To complete their work on the boundary survey, Lambart, Wheeler and Cautley all needed a large baseline to connect with their survey stations. The Geodetic Survey had established a baseline at Yellowhead Lake at the southern end of the network, and needed a similar one at the northern end. Cautley had started at the longitude pier near Pouce Coupe and also required a similar station near the intersection with the Rocky Mountains. Wheeler and Campbell would use this baseline for their triangulation network, connecting from the two ends of the base to the stations visible around them. Initially the baseline was planned for the Jarvis Pass area. However, when Lambart found that the intersection of the 120th meridian and the divide was much farther south, the location for the baseline was moved to Sheep Creek valley, a few kilometres north of the intersection. Arrangements were also made for F.A. McDiarmid, who had established the longitude pier, to come to Sheep Creek to determine the location of the 120th meridian along the baseline.

Lambart had the responsibility for establishing the baseline, and he started this project on the last day of June:

The end of the month sees the work well commenced as today, although leisurely, saw the new base decided upon and the expansion laid out in fine style. It's good working with Hans, he sees points that I might've missed & I hope vice versa. There is not a bit of cutting on the whole

Lambart's crew surveying at a geodetic station. e011166437, Library and Archives Canada

thing and the length judged to be about 3 miles [4.8 kilometres]. Another mile or so could've been introduced without any cutting being necessary.

On July 1, crew members set signals at two nearby geodetic stations; Hans and his packers put in the east base; and Lambart and his packers put in the west base. "Thus in two days we had the base located and in readiness for observing," Lambart wrote. Then it was time to celebrate the holiday: "In the evening both parties got to games, quoits and the football until late."

On July 2, Lambart and his crew, including Arthur, climbed to a station, "this being the commencement of the observing for the summer." Felix and Hans's packers went down Sheep Creek to bring up the sup-

　　　SURVEYING THE 120TH MERIDIAN AND THE GREAT DIVIDE

Pang geodetic station. Lambart's caption: "The signal has been erected in a temporary cairn, while the main station is occupied by the theodolite [transit], since another engineer is sighting on the signal from a distant mountain. When the work is completed the signal and large cairn will be rebuilt on the station marker." e011166438, Library and Archives Canada

plies from the cache Hughes had constructed when he brought the goods in May. The following day, Lambart and Arthur climbed to the geodetic station on Intersection Mountain. Hans and his crew selected one route, while Lambart picked a route that involved a "little rock experience… for Arthur & I am sure will be good for him. He did well in spite of a feeling of nervousness, as to any beginner the positions we were in were certainly terrifying." On July 4, Lambart surveyed from the second Geodetic Survey station nearby: "I never remember spending a more perfect day on top of a mountain. It was hot and no wind blowing and the horizon as clear as could be."

The next day, Lambart travelled to McBrien Lake, where he "found my old cache of last year more or less in ruins. The porcupines had dropped the bag of provisions on the ground which had been pretty thoroughly gone through." On July 6, while travelling around the area, "a letter was left on a stick on the trail at Wapoon Lake for Cautley, giving him all the information we could and acquainting him with our movements and the stations and the baseline operations at the Sheep Creek pass which we told him was seven hours distant from Lake McBrien."

After a day of preparations, Lambart departed on July 8, heading to Robson Pass to participate in the ACC camp. On the first day, he met A.J. Campbell and spent forty minutes visiting with him. The next day, Lambart decided to try a different route. "Today we start on our unknown journey up the limestone valley to seek out the new way to avoid the climbing of Big Shale Mountain." The men wound up in a difficult section of the Jackpine River.

> We got on a steep side hill and one of the horses took a headlong plunge 40 feet [12.2 metres] to the river below where it was running in a rapid. It happened to light on its pack twice or it certainly would have been killed. However, good fortune was with us this time and the poor brute made the other side alright and got out. We tried to make another ford 100 yards [91 metres] above, the approach to which was a steep bank down to the edge of the stream. Here I volunteered to try first and with a rope around me I started across the stiff running water. A quarter of the way across Tony stepped on a shelving smooth rock and I went off, the rope tightened and I was helpless. I was above the horse & a little afraid at first of getting under his feet, he got his feet and went across fine while I was dragged helplessly back to shore.

By the evening of the next day, they were back on their route and at a campground they had previously used. "This means that it just cost us in experience 15 hours hard work to avoid the Big Shale, a goodly price to pay." Their difficulties with the horses continued the next day while travelling along the river. "We had the misfortune to have the horse with the stationery and the camera go over the bank getting the contents of both in a shocking condition. We opened the boxes immediately and poured out the water. In camp tonight we are hard at the drying process."

On July 12, Lambart met John Kihl and learned about the surveying he had completed. The next day, Lambart's crew reached Carcajou campground along the Smoky River. Before leaving the next morning, Lambart made a diary entry.

> We were anxious about our horses lest they would play us the old prank here of scattering away down the Smoky or away back to the head of the river. Felix and Ed were up during the night with them and succeeded in holding them near camp until the morning. We are delayed with the finding of a few this morning and having to make a special trip to find a crossing of the Carcajou which is high. I don't know what is before us this morning, we have to make our way all the way through on the west side.

Lambart reached Robson Pass that evening, and the next day travelled down to Robson Station, where he took the train to Jasper. F.A. McDiarmid met him at the station, and the two men discussed the baseline at Sheep Creek. On July 16, "McDiarmid and party went east to mouth of Snake Indian 20 miles down the track on a freight train. The horses light left late in the morning with George Plante." Since Hughes had already brought in supplies, the pack train had to take only items needed for the journey.

On July 19, Lambart and his son returned to Robson Pass, and they spent the next two weeks at the ACC camp. On July 23, two groups climbed Gendarme Mountain, northwest of Robson Pass. At Wheeler's request, Lambart led the group with Don and Phyllis Munday, the famous Vancouver mountaineers. Arthur was a member of the other climbing group, but "about halfway up the slope we left Arthur behind until we should return, he being very tired and his legs aching." During the camp, Conrad Kain successfully guided two groups to the summit of Mount Robson. Lambart was in the second one, which included Phyllis Munday and Annette Buck from New York, the first two women to climb the highest mountain in the Canadian Rockies.

CEREMONY AT MOUNT ROBSON

Lambart helped erect the monument commemorating the boundary survey, and he participated in the ceremony on July 31, representing the Geodetic Survey.

Other Dominion government officials who attended included: J.N. Wallace, the Dominion government's commissioner from 1913 to 1915 and director of levels; and F.H.H. Williamson, deputy commissioner of the national parks. BC was represented by its surveyor general, J.E. Umbach; the deputy minister of lands, G.R. Naden; and BC's boundary commissioner, A.O. Wheeler. P.N. Johnson, director of surveys, represented Alberta. Representatives from the Canadian National and Canadian Pacific railways also attended. A.J. Campbell arrived at Robson Pass on the day of the ceremony, but it appears that it was after the event. In his diary entry for July 31, Lambart wrote:

> A day of ceremonials. In the morning the unveiling of the monument was interrupted by a slight shower when I was intensely relieved by this, getting out of making a little speech. It came off very successfully. Mrs. Campbell doing the unveiling. Movie and camera at work and short addresses from Umbach, Naden, Wallace and Wheeler.

The Boundary Commission report provides more detail.

> The ceremony was an impressive one. The towering magnificence of the mighty snow-clad mass of Mt. Robson, rising directly above, gave solemnity and dignity to the scene. The surroundings were certainly unique; the great mountain whose crest, golden in the fitful sunshine and then obscured by passing clouds, was a full mile above us, close by the blue waters of Berg and Adolphus Lakes, and all around wide-spreading snowfields and tumbling icefalls between precipitous rock ramparts, the crowd of picturesquely garbed mountain climbers, men and women, assembled about the monument, which occupied the centre of a broad bare shingle flat, brought vividly home to us the magnitude of the works of nature at their origin and the wonderful heritage we possess in this mountainous wilderness of unsurpassed scenic grandeur.

The report also noted that the monument had an inscription plate on the Alberta side with Deville's name on it.

Left to right: H.F. Lambart, G.R. Naden (BC deputy minister of lands), J.E. Umbach (BC surveyor general), A.O. Wheeler, J.N. Wallace (Dominion director of levels and boundary commissioner, 1913 to 1915). PA-202781, Library and Archives Canada

Sheep Creek baseline. Lambart wrote: "The rock cairn marks the intersection of the 120th meridian boundary with the Sheep Creek base line at the northern end of the boundary triangulation. Twelve feet [3.6 metres] to the right is the signal erected over West Base geodetic station." e011166439, Library and Archives Canada

As the ACC camp concluded, Lambart was eager to get back to surveying, but August 3 was a Sunday, it was raining, and the crew spent all of the next day searching for the horses. On the morning of August 5, the final two horses were located, and the crew left after lunch. Two days later, Lambart met John Kihl and his crew at the Resthaven camp. "He had most excellent news to report. He had finished on Curly and was on his way out. I told him that after he finished with his return climb on Carcajou we had Mumm, Lynx, and Calumet left to do. I was glad now that the stations were there to do as the season had closed extremely quickly." The following day, August 8, he met Hans Kihl. "He had also finished up all his work and gave me much news of McDiarmid and his crew and all their troubles and tribulations. Anxious to have a talk together, he very kindly consented to come back to 'mail by heck' camp. A great talk and gathering about the evening bonfire."

On August 9, about an hour after crossing Big Shale Pass, Lambart met McDiarmid at a camp.

> Mac finished up splendidly and completed azimuth, latitude and longitude observations as well as measuring the base four times in record time. The azimuth mark he placed on the south side. The 120th Meridian came about 8 feet [2.4 metres] inside the base from West base. Cautley is reported camped near the base. The line also came only about 200 feet [60 metres] from Arthur [a geodetic station]. This short camp places us behind one day in our schedule as we had hoped to make the Sheep Creek pass on the 10th.

On August 10, Lambart took a short trip to the McDonald camp.

From Sheep Creek valley Lambart took this photo showing Intersection Mountain (Haig) and Casket Mountain (Coffin). e011166441, Library and Archives Canada

Campbell had caught up and was in camp. We had a good wash in the lake. Took photographs of the collection of gear left behind by No. 3, a complete set of musical instruments and a set of golf sticks. The evening in the old campground was enjoyable with not a care in the world, all the work being now completed with the exception of the points at the pass (Robson) which I take as a matter of course will be finished up by Hans and John who are anyway just looking for something to keep them going a little while longer.

Lambart reached Sheep Creek the next day; Campbell had arrived about thirty minutes earlier. August 12 was spent cleaning up around the site.

Campbell departed the next morning.

It's a bright day and Campbell moves out to join his party to the north and last seen occupying Minnes station, one of our points of last year and now cut out by the location of the base on the Sheep instead of on the Porcupine east of the Jarvis Pass. He had six head and Tommy Plante was along.

Before beginning his return journey, Lambart decided to take his crew and go on a sheep hunting trip. In his diary entry for August 16, he described "a day of great regrets to me and which I want to forget on account of the regrettable slaughtering of 5 and perhaps 6 sheep which came in between me and the big fellow I was after." Lambart depicted the scene in detail and then wrote: "On standing up it was terrible to see six beautiful animals wounded and in their agony trying

Felix Plante (left) and unidentified man entertaining at camp. e011205428_s1, Library and Archives Canada

to get away from the dreadful enemy whom perhaps [they] had never seen before." Lambart immediately felt remorse for his actions: "I was feeling terribly over the accident and was really in a dilemma what was the best course of action.… With great regret we decided to take what we could, by packing back into camp and just letting the rest go." At the conclusion of his entry for August 16, Lambart wrote: "So ends the day I shall never forget and ever deeply regret." Lambart and the men returned to the camp at Sheep Creek on August 18.

Lambart spent August 19 around the area. "Arthur as usual standing about with his hands in his pockets." He described the scene at the baseline:

Christy, Arthur and I walked up the pass to west base and examined Cautley's work there, the 120th Meridian making a remarkable intersection of the baseline not more than 8 feet [2.4 metres] within the base from west base. The intersection had been marked by a sunken tube with copper top, and to the south about 8 feet distant from the intersection, a 10 foot cairn [3 metres] had been erected, surmounted by a pointed 8 foot pole. The vista was cut out and everything had been left in shipshape order.

Although August 20 was a "miserable, cold rainy day," Lambart's crew started their return journey. En route that day, they stopped at a cache they had made exactly a year earlier. After taking a few items, they left the rest for Curly Phillips. They camped that night at Lower McDonald camp. "Cautley had been here before us a few days and has left some welcome piles of cut wood." The men remained at this camp for a second night because of the rain. While travelling on August 22, "we came upon 6 lovely caribou and had the pleasure of watching them for some time in the open before they decided to get away to cover."

Before leaving camp on August 23, "we debated the question of ditching the stove, tables and chairs. With the exception of the stove, Felix expressed a desire to have [them] so I gave them to him and we took them along. My total contribution to the Plante family was: two tables, two chairs, two axes, 10 pounds [4.5 kilograms] sugar, 6 pounds [2.7 kilograms] of apples, 20 pounds [9 kilograms] other provisions."

It rained on August 24. "Felix did well on the many crossings of Bess Creek which was a raging torrent and had never been so high whenever we had occasion to cross it. It was nearly a swim at the lower end where all the streams came together, but we managed it without getting anything wet in the boxes." The next day, Lambart's crew reached the camp on Adolphus Lake, where they found Hans Kihl. "Hans had still Mumm and Lynx on his hands and had been in this camp about 10 days waiting for decent weather to finish. The instrument was cached on Lynx."

Despite relentless rain, Lambart packed out to Robson Station on August 26:

> The waters are very high and the journey today was marked by frequent wadings, where the water in many places had come

over the trail…. Again the little pang at the close of another season in the field and especially this one which has been an exceedingly eventful one for many of us…. On the whole the season has gone by with personnel which have been eminently satisfactory.

While travelling to Jasper that evening, "Constable Sinclair got on the train with us and it gave me an opportunity of confessing to the story of the accidental killing of the sheep. He took all down verbatim and was good enough to take it as an accident pure and simple."

On August 28, Lambart left for Ottawa in the evening.

> With the valued assistance of Isaac Plante, I got all our outfit in good shape for the express…. All that was left behind was the instrument boxes and personal stuff of the two parties still in the field (Hans and John)…. I wired Felix yesterday at Red Pass to see Sinclair at Lucerne when he went through. Presented Conrad with a pen…. We parted company with the Brewsters at the station after three years of most pleasant business relations.

The Geodetic Survey's triangulation network provided great assistance to the boundary survey, especially for Wheeler and Campbell's surveying. It was Lambart's first experience in the Rocky Mountains, and he provided many prosaic descriptions of the scenery, as well as the flora and fauna. In his 1924 annual report, Lambart wrote:

> The country by this triangulation net is one of the most magnificent stretches of the Rocky Mountains. Wonderful scenery, with rugged peaks, large ice fields, a profusion of alpine flora and grasses, sulphur springs and generally brilliant weather during June, July and August, make this area one of surpassing attraction. No

Horses packing monument material to the summit of Phillipps Pass. e011205466, Library and Archives Canada

SURVEYING THE 120TH MERIDIAN AND THE GREAT DIVIDE

minerals were noted, but game is quite plentiful, a few deer, many caribou, with an occasional grizzly bear and mountain goat which become more numerous to the north of Big Shale Mountain.

R.W. Cautley

Cautley's main objective was to complete the survey of the southern portion of the 120th meridian. But for the beginning of the field season, he had to undertake a small project. The survey of Phillipps Pass remained from the 1914 survey of the Crowsnest Pass area, and at the 1921 conference, a decision was made to run the boundary line through the middle of the pass. The pass was close to a rail line and there was a road through it, so there was easy access, and Cautley would be able to complete the survey in the spring while his location on the 120th meridian was still inaccessible.

Cautley and one assistant arrived in Crowsnest Pass on May 6, and spent the first day examining the pass and arranging for the horses and packer. The next day he located a source of gravel nearby. "Loaded wagon and went to summit Phillipps Pass. Pack horses met us at the pass and we took monument material up to 18-F and put in top of monument. The finest climb I ever saw horses make. 3300 feet [1005 metres] climbing."

During the next week, Cautley established four monuments across the pass. The most northerly one was a bolt and cairn on Mount Tecumseh. He surveyed the boundary line and made a survey of the small Phillipps Lake. Generally, the work proceeded smoothly. However, on May 8, "horses had accident coming down from 18-F and bent shoveling boards badly," and on May 13, "cut out and chained very steep course between 14-F and 16-F. Broke chain doing so. 5000 foot [1525 metres] climbing." Cautley completed the survey of Phillipps Pass by May 14, and the next day he arrived in Edmonton. He finished the office work for this project before leaving for the 120th meridian.

On June 1, Wheeler came to Edmonton, and the two men discussed plans for the summer. Cautley and his crew departed for Grande Prairie on the afternoon of June 5, arriving the next day. On June 7, the men

Phillipps Pass and one of the stations bisecting the small lake there. e011205431, Library and Archives Canada

travelled to Rio Grande and camped on MacDonell's property. The packers were there, and Cautley found that the horses were in excellent condition. After a Sunday in camp, the crew started for the Mount Torrens area on June 9. As usual, there were problems on the first day. "Moved 15 miles [24 kilometres] and camped at Belcourts. Got away at 10 AM. Buck bucked pack off, broke saddle, and several delays. Cook and Bob foolishly went 6 miles [9.6 kilometres] beyond camp." The next day they reached the Wapiti River, where they "found water very high and ford impassable."

Cautley's crew in Sheep Creek valley. e011205468, Library and Archives Canada

It took two days to get the equipment across the river. They got the canvas boat out, but it "was found to lack 3 ribs. Made ribs of red willow. Got rope across just below where we put horses in. Packed cache up there and put 22 loads across in canvas boat—about 340 pounds [154 kilograms] each trip. Swam 4 horses and moved stuff high above bar on which it was landed. Everybody did well."

On June 12, the crew

finished ferrying loads across river by noon. Two horses went back to Calliou Lake. Loaded all camp outfit on horses and carried 30.00 [chains, 600 metres] downstream to old log cache and there camped 200 yards [183 metres] from camp reached

Tuesday. Loaded spare cache on horses and sent 5 miles [8 kilometres] up to the nearest feed. Two horses nearly drowned swimming river. Saw moose cow and two calves trying to cross river. Very late stopping work but no feed here.

The crew moved south of the Wapiti on June 14. One of Cautley's tripods was badly damaged in a packing accident. Cautley "spent half the night repairing. Horses tired out with heavy packs up big Wapiti hill and wet trails." The next day "Mr. Gorman and self spent all morning completing repairs to tripod which is now as good as ever but 7½ inches [18 centimetres] shorter."

On June 16, "self and packers went back 9 miles [14 kilometres] and brought up the whole 4100 pounds [1860 kilograms] left at the cache on Thursday night [June 12]. Mr. Gorman and rest of party made a bear-proof cache in which to have the second load and rebuild bridges on the trail ahead."

The crew reached the Narraway River by June 18; the river was high after a couple of days of rain. The next morning

> self and head packer walk down to the ford 2 miles [3.2 kilometres] away and thought we could get the outfit across, so we went back and packed up. On reaching the ford I went across on my saddle mare and got back safely although the water was over the seat of my saddle in one place and the strength of the current was such that I expected her to lose her feet both going and coming. It was not a safe risk for the big horses and no chance at all the small ones so we camped at the river.

Heavy rain and sleet continued the next day. "Contrary to my expectation I found a place where we could use the canvas boat and swim the horses, so sent Pascal back with one pack horse to get it from Wapiti, where I had cached it because I did not expect to need it and could just move the outfit in two trips by heavy packing without it. Water fell about 3" [7.5 centimetres]."

On June 21, "water had fallen 6½ inches [16 centimetres] since my Thursday trial and we succeeded in moving camp 9 miles [14 kilometres] to south side of Torrens. Crossing the river was anxious work, the water being at about the limit of possible fording, but we lost nothing." Cautley was now at the location where his 1924 surveying would begin. June 22 was Sunday, but while the men rested, Cautley went out to gather information. The next day, "self, Mr. Gorman and Nicholson climbed to 61-5, Torrens, and 61-4 [the station along the 120th meridian on Mount Torrens] and planned triangles. Set signal at 61-5." Cautley was ready to begin surveying.

Although the purpose of the survey, the production of the 120th meridian line, remained the same, the surveying methods used in 1924 had changed considerably from previous years. The Boundary Commission report explained the details.

> In order to fulfill the special objectives of the survey it was necessary to conduct three separate and distinct operations. First, the 120th meridian itself had to be surveyed and monumented as a straight-line boundary. Second, a triangulation net had to be carried forward in the vicinity of the 120th meridian, so that the distances between monuments could be ascertained. Third, a phototopographic survey of the confused mountain topography of the region was required. The first two were undertaken by Mr. Cautley's division, and the third by Mr. Wheeler's.

The report stated that the triangulation net started at a base "measured between Monument 62-3 and 62-6 on the meridian itself" in 1922 near the Narraway River. It continued:

> From the above base the net was gradually expanded until the sides of the triangles were about seven miles [11 kilometres] in length. Finally, the net was closed on a triangulation base in the valley of Sheep Creek, surveyed in August, 1924

by Mr F.A. McDiarmid of the Geodetic Survey of Canada.

During 1924, Cautley was able to connect his surveying with some of the stations of Lambart's geodetic triangulation, facilitating his work. The report also noted:

> The survey of any straight line through really mountainous country is a most unusual proceeding, and involves considerable difficulty. There are many mountain areas in the Rockies which are so rugged in character that of all the infinite number of potential straight lines which might be laid down only an insignificant number could, in fact, be surveyed at all. Fortunately, the mountains along the 120th meridian are, for the most part, of a secondary character and presented no insuperable difficulties.

The report explained the production of the 120th meridian in 1924:

> Starting at Monument 61-4 the line was projected southerly to the next sharply defined mountain ridge that commanded a clear view along the meridian in both directions, and thenceforward, from ridge to ridge.
>
> Monuments consisting of specially marked bolts cemented in rock, over which were erected well-built cairns containing from three and a half to seven tons of rock, were established on all the dominant ridges occupied as stations.... Altogether fifteen monuments were established in a total of 33.9 miles [54 kilometres], of which all except two are of the bolt and cairn type.

The two exceptions were in the Kakwa and Sheep Creek valleys, "where a certain amount of line was cut as a means towards finding the monuments estab-

lished in these valleys." Sheep Creek valley was especially important, since the baseline used by all three survey parties was established there.

Although it was of secondary importance, elevations along the boundary line had significance for mapping the topography. The Boundary Commission report stated:

> During the years that Mr. Cautley's division was engaged in carrying levels 116 miles [187 kilometres] south from Fort St. John datum to the intersection, Mr. Wheeler's division was carrying trigonometrical levels north from Yellowhead Pass railway datum to the same point, a distance of about 130 miles [209 kilometres] over very rough mountains. Where the work of the two divisions was joined up, namely at the intersection of the 120th meridian with the summit of the Rockies, there was a difference of 12 feet [3.7 metres] between them.

This discrepancy was small enough for the distance measured and terrain covered to enable accurate mapping of the topography.

The report described how the main difficulty in surveying the 120th meridian

> was due to the fact that most of the stations were so high above the valleys, and involved so much hard climbing, that it was found impracticable to use the large transit for producing, and that the lighter transits, while excellent angle-reading instruments, proved insufficient for first-class line production. For the above cause it was found necessary to observe for azimuth as much as possible, and to depend on astronomic control for the correction of the line from time to time.

Since Cautley did not have to chain distances or construct concrete monuments, and had to do only minimal clearing of line, he had a smaller crew than in previous years.

On June 24, Cautley produced the 120th meridian line to the next station, 60-3, on the summit of a ridge, while Gorman read angles at that station. Cautley recorded that "great difficulty was experienced in getting signals." The line had to be produced exactly, and the station was over 5 kilometres away. Gorman surveyed from a triangulation station the next day while Cautley, with the assistance of two crew members, "packed the big instrument and my fly camping outfit up to 60-3 where I stayed overnight while the men went back. I succeeded in getting one time and two azimuth observations." The surveying proceeded efficiently, with both Gorman and Cautley taking small groups of men to assist them with the work.

On June 27, Cautley wrote in his diary: "The packers arrived with everything having made an excellent trip.… It is a great relief to have everything safely on this side of Narraway River." The next day, Cautley and his crew finished their work on Mount Torrens. "Self and four men went up to Torrens summit and read final angles, building up cairn again to its full original size. Also took photographs. On way back read final angles at 61-4 and also built up cairn again with large signal." Unfortunately, Cautley "wrenched back badly."

Meanwhile, Gorman "established station G within sight of 16th [baseline] and read on stations already established." The Boundary Commission report observed that this "was the only previously surveyed line with which connection was made during the season of 1924."

The crew moved camp to the east fork of the Narraway River on the last day of the month. Cautley's sore back kept him in camp for a few days, while Gorman worked on making a good surveying connection with the 16th baseline. During the end of the first week in July, Cautley and Gorman produced the meridian line to the next two stations, 60-2 and 60-1, which were a few chains east of Mount Gorman. The two men climbed to the summit of Mount Gorman on July 8 and surveyed from a station they established there. Two days later, Cautley returned to the mountain. "Self and Ed climbed to 60-1 and succeeded in putting in bolt, reading all angles and erecting first-class cairn, but I never suffered so much from cold on a winter's survey."

The surveying proceeded uneventfully through July, producing the meridian line and surveying from stations on the triangulation network. When the nights were clear, Cautley and Gorman made observations. A.J. Campbell arrived at their camp on the evening of July 26, departing the next morning for the ceremony at Robson Pass and taking their mail with him.

Cautley wrote a letter to the surveyor general on July 28 describing his progress.

> Monday, 23rd June, up to the present date the party have practically completed 18 miles [29 kilometres] of the 32 miles [51 kilometres] from Torrens to Haig, besides which some forward work has been accomplished. From now on I expect to get along faster, because we have now tied on to Mr. Lambart's work and shall be able to use many of his cairns; which does away with the necessity for establishing forward cairns—at least on control points of the triangulation—for ourselves.

He also noted:

> Fortunately it is an easy country to make a triangulation survey in; there are a great many mountains from 7500 to 8000 feet [2286 to 2438 metres] to choose from and we have succeeded in getting good triangles. Tying onto the 16th base line proved to be somewhat difficult, because the end of the line occurs on a very steep wooded hillside facing north; however a satisfactory tie was secured.

Cautley surveyed from one of Lambart's stations on July 30. "Self climbed Edward and got a large circle of azimuths and observations. We saw a bull moose, three caribou and 12 different bands of goats—one of them with 19, all nannies and kids. I suggest maternity or crèche as name for mountain [today called La Crèche Mountain]."

On the last day of the month, the crew moved to a camp about 2.5 kilometres south of Cecilia Lake. From this location, during the first days of August, Cautley and Gorman continued producing the 120th meridian and surveyed from Lambart's stations in the vicinity. Rain prevented work on two days.

Cautley moved camp to the Sheep Creek valley on August 6. His timing was impeccable, for McDiarmid was there and had just finished measuring the baseline. The Boundary Commission report described McDiarmid's work at the baseline and the new technology that he was using.

> Sheep Creek base is just over three miles [5 kilometres] in length, and was measured by the most precise methods known to the profession of surveying.... The extremities of the base are marked by bolts set in concrete....
>
> Mr. McDiarmid also established the latitude and longitude of the easterly extremity of the base by a series of precise astronomical observations....
>
> In regard to the longitude observations obtained by Mr. McDiarmid, it is interesting to note the revolution that has taken place in the practice of astronomical determination of longitude since Mr. McDiarmid established the longitude pier at Pouce Coupe in 1917. At that time it was impossible to make an astronomical determination of longitude, with anything approaching the degree of precision required for the establishment of an interprovincial boundary, at any place that was not directly connected by an ordinary telegraph line with some point of previously determined longitude. In 1924, owing to the development of radio-telegraphy, it was possible to make such a determination at any point whatsoever, leaving out of consideration what are known as "dead" areas. It would be difficult to exaggerate the immense importance of radio-telegraphy to those who undertake the mapping of unexplored parts of the world. In the case of the Sheep Creek determinations, telegraphic time

signals were received direct from Annapolis, USA [Maryland] and were very clear.

The worldwide use of radio signals to determine longitude was one of the most important technological developments for surveying in the 1920s.

The discrepancy between Cautley's meridian line at the Sheep Creek baseline and McDiarmid's determination of the 120th meridian was 5.8 metres over approximately 185 kilometres, plus the survey of approximately 9 kilometres from the Pouce Coupe longitude pier to the meridian. Cautley had made an excellent survey of the 120th meridian.

Cautley made observations on the stars that night and spent the next evening visiting McDiarmid. On August 8, Cautley surveyed from the eastern base and also went to examine Sheep Creek pass, one of the potential passes to survey after completing the meridian line. He found that this pass was entirely in British Columbia.

The next day, Cautley produced the 120th meridian line about 5 kilometres from Sheep Creek to its intersection with the Great Divide. It was on a mountain initially called Haig, but later named Intersection. Lambart had established two stations there, one of them less than 20 metres from 56-0, Cautley's final station. Unfortunately, the 120th meridian did not cross Intersection Mountain at a prominent location, as described in the Boundary Commission report.

> Monument 56-0 marks a spot which is shown on every atlas of Canada, and which your Commissioners have been striving towards for years. It is a matter for regret that, in spite of what would seem from the above description to be a very commanding position, Monument 56-0 is really very inconspicuous. This is largely due to the fact that the ridge of Intersection Mountain lies nearly north and south, with various lateral ridges which cut off the view from any point in Sheep Creek valley that is near the boundary.

Cautley surveyed from his station and Lambart's nearby Haig West. He also constructed a large cairn at

56-0. On August 11, he and Gorman each took a small crew on a fly camp to survey at two of Lambart's stations that were farther away. During the first night, Cautley wrote, "a caribou bull came within 50 yards [46 metres] of the tent and bellowed at us, apparently seeking a fight." He surveyed two additional stations before returning to the main camp. Cautley completed his survey of the 120th meridian on August 14.

> Self climbed to 56-1 [the first station north of the intersection], put in bolt, read, built cairn, and photographed it. In p.m. read elevations at 56-2 [Sheep Creek], made final measurements base line west, built an enormous cairn at 56-2, photographed and cut line through bush to south.

Now Cautley was ready to begin the second part of the season's work: the survey of Robson and Miette passes. The crew departed on the afternoon of August 15. Three days later, Wheeler's head packer, who was also going to Robson Pass, joined them. On August 21, while travelling along the Jackpine, four of the horses went into the river and got their packs wet. August 22 was their first sunny day since August 13, and they travelled over 20 kilometres across Jackpine and Bess passes and down to the Smoky River. Because of the rain and other factors, Cautley did not reach Robson Pass until August 25. There he found Wheeler waiting. "Everybody and everything very wet."

Cautley spent about two weeks surveying the pass. It was the only one to have a distinct station at the lowest part of a pass. The Boundary Commission report stated: "By directions of the Governments, Monument 1-U is in the nature of a memorial monument commemorating the conclusion of the Interprovincial Boundary Survey." Cautley and Wheeler spent August 26 selecting the first stations above the pass (2-U and 3-U). The next day, the two men went up the slope of Mumm Peak and chose the site for 4-U. On August 28, they went up the opposite side. "Self and Mr. Wheeler climbed Ptarmigan Ridge and determined watershed from 3-U up, putting in bolt 5-U." Meanwhile, Gorman and the rest of the crew cut the line and started constructing the monuments. He and

Cautley took this photograph and drew the projection of the 120th meridian across Sheep Creek valley to Intersection Mountain where it meets the Great Divide. e01125433, Library and Archives Canada

Cautley also established a triangulation base, and the crew put in a big signal at the east end. On August 30, the packers left with the first load for Miette Pass.

Wheeler departed for Sheep Creek on September 1, while Cautley and Gorman worked on the triangulation base. Rain limited work the next two days. September 4 was a

> beautiful day. Mr. Gorman climbed to summit Ptarmigan and occupied cairn station there. Self climbed to bolt 4A-U and occupied it. Small space on top and rotten rock made it necessary to read six circles of azimuths. Photographed and built good cairn. Came down to 4-U. Photographed concrete monument and read angles and observations. [4A-U was an additional station on the slopes of Mumm Peak. 6-U is a station on the summit of Mumm Peak.]

With one more day of nice weather, Cautley was able to complete the survey of Robson Pass, and the packers returned from Miette Pass.

While his crew started towards Miette Pass, Cautley took two pack horses and went to Robson Station on September 6. Leaving his horses at Hargreave's, Cautley "went on by night train to Yellowhead to arrange with Park Warden Bigley about taking care of freight, but was taken on to Yellowhead Station 4½ miles [7 km] east of summit because I found out too late train did not stop at summit. Slept in ditch." The next day, he "wrote 3 letters to Bigley & left one at Yellowhead Station, one at Lucerne, and one with big canvas which I threw off train at Yellowhead summit in front of his velocipede shed. Returned to Robson and worked all day helping shoe horses. Left Hargreave's at 4:35 PM and arrived Robson summit at 11 PM." On September 8, Cautley travelled 33 kilometres and rejoined his crew after dark. "One of my pack horses took a swim in Moose River with my bed."

The crew reached Miette Pass on September 10, encountering some snow on the upper portion of the trail. Cautley made a preliminary examination of the pass. The Boundary Commission report described the location: "Miette Pass is a gap of about three miles [4.8

kilometres] between high mountains of the main divide, in which there are three distinct passages separated by two ridges about 1100 feet [335 metres] above the centre pass." The centre pass is the lowest, while the north is the highest. The total distance Cautley measured from the peaks along the Great Divide at both ends of Miette Pass was almost 8 kilometres.

The next day, the crew located a source of gravel about 2.5 kilometres away, while Cautley and Gorman "located and marked watershed through centre and north passages—about 3 miles [4.8 kilometres]—so that Mr. Gorman can have preliminary survey made during my absence. Tomorrow I am going into Yellowhead and Jasper with packers in case freight has not arrived or has not been left at the summit." On September 12, Cautley "went in to Yellowhead summit, thence to Lucerne, and thence to Jasper taking pattern of such horses' feet as absolutely cannot go on without shoes." He found the freight at Yellowhead summit. The horses got shod and the freight was brought back to camp by September 16.

Wheeler had not yet arrived, but Cautley needed to finish surveying the pass as soon as possible. Although it was wide, the watershed through the pass was easy to determine. From the lowest point in the centre pass, there were four stations on the north side and five on the south. When the horses arrived in camp, the crew "proceeded at once, without unpacking, to distribute cement to 1-T, 2-T, 3-T and 5-T, also got gravel enough up to filled tops of 1-T and 2-T so that they can be built tomorrow." Work continued the next day. It snowed that night. The crew tried to work on some stations on September 18, but "driving wind and snow made it impossible to stay on the exposed ridge, let alone do any work and the attempt had to be abandoned."

This was the beginning of several days of snow. But the crew had to try to complete the work, for this was the end of the boundary survey. In his diary entry for September 20, Cautley wrote:

> Still snowing and blowing. Made four loads of dry wood, packed them up to 4-T and 6-T respectively with big tub and self, Sam and Ed succeeded in building both monuments by use of hot water. Heated

rocks to put around monuments and covered all with canvas. Mr. Gorman and Tom precise chaining baseline and distance from 1-T to 2-T. Packers took gravel and cement up to 7-T.

By September 22, the snow was 40 centimetres deep on the valley floor.

While Cautley's crew was working in snowy Miette Pass, Édouard-Gaston Deville died. The Dominion surveyor general for thirty-nine years, he had lived almost to the completion of the survey, and knew that it would be successfully finished. In recognition of his role on the boundary survey, his name was inscribed on the Alberta side of the memorial monument at Robson Pass. Commenting on the inscription, the Boundary Commission report described Deville as "a man and a scientist to whom Canada owes most largely her magnificent system of land surveys, and also the introduction of the method of photography, a method so well suited to her mountain areas and so successfully carried on in mapping them. It is fitting that his name should be on record at a place where their grandeur reaches a climax."

On September 23, Cautley wrote:

Snowing all night and until noon when it cleared for about two hours. Went up to 4-T to take forms off. By the time I got up there a fierce blizzard had blown up and I actually froze my fingers while taking off forms although the temperature cannot very well have been lower than 25°F. Brought forms into camp. Snow 36 inches [0.9 metres] deep on higher slopes. Sam brought in forms from 6-T and had similar experience.

His diary entry the next day described similar conditions:

Snowed during night and until noon. Sam and Tom built 5-T. Mr. Gorman and two men dug horse trail up towards 7-T and packers drove all the horses over it to beat it down. Self and Bob went out to 6-T and packed in shoveling board and tools as it is impossible to get horses over there, snow being from 24 inches to 40 inches [0.6 to 1 metre] deep. Started snowing again at 6 PM and snowed all evening.

On September 25, "Mr. Wheeler and outfit came into camp from Colonel Pass in a wild snowstorm; we all turned in and dug out a camp for them."

The weather finally cleared on September 26.

There is 21 inches of snow [53 centimetres] this morning. Cleared up in a.m. I explained the survey as made to Mr. Wheeler and in pm he, I, and Mr. Gorman went up to 7-T and put in a bolt at 9-T. Horses packed wood up to 7-T and top of same was filled. Sam and Tom built 3-T. Mr. Wheeler and I sent all our horses out to break trail through south passage.

Wheeler departed the next afternoon, and during the following four days, Cautley's crew finished most of the surveying of Miette Pass. On October 2, there was a

heavy snowstorm. Moved 5 miles [8 kilometres] from camp in centre pass to 1½ miles [2.4 kilometres] on Alberta side of south pass. It snowed two inches [5 centimetres] while we were packing up & continued all day but I was afraid of being storm bound & anxious to get the horses over the high ground between centre and south passes so moved anyhow. I had hoped to finish the survey in the south pass also but it was out of the question.

October 3 was a landmark day:

Light snow with a driving wind & very cold. Self read at west end south pass base and at 11-T south and at 11-T. Mr. Gorman precise chained & checked south pass base and distance from 11-T to 11-T south. The weather made this day's work very severe on all hands. This day saw the completion of the field work of the Interprovincial Boundary Commission after 12 seasons work.

In his government report, Wheeler described the survey of Miette Pass.

> Most of the time the temp was below freezing and there were high winds. Under the circumstances it was necessary to adopt arduous methods: horse trails had to be shovelled out to the higher monument sites; dry wood and tubs in which to melt snow had to be packed up; the concrete had to be made with hot water so as to overcome the frozen condition of the gravel; and hot rocks laid under canvas around the green concrete to permit it to set. Angles read at exposed stations above timberline under such weather conditions are naturally regarded with suspicion, but we were relieved to find our triangles closed within perfectly normal limits.
>   Much credit is due to Mr. Cautley for carrying on and completing the work under such unfavourable conditions, and it is due to his well-systematized methods and dogged perseverance against severe hardship that it was completed.

The next day, the crew reached Yellowhead Pass, travelling through deep snow for much of the day. The crew members, except for Cautley, Gorman and the packers, departed by train the following evening. Cautley and his group arrived in Jasper on October 6. The packers had to get back to the Peace River region, so Cautley paid their expenses and gave them a week's wages. They finished their work and departed on October 9. The next day, the horses were sold by auction, and Cautley left Jasper by night train, arriving in Edmonton on October 11.

Cautley in Miette Pass. e011205434, Library and Archives Canada

N.C. Stewart is standing at the intersection of the 120th Meridian with the 60th parallel. This is the northernmost point of the Alberta-British Columbia boundary where it adjoins the southern boundary of the Northwest Territories. Stewart was an assistant on Wheeler's 1924 crew and he was British Columbia's boundary commissioner from 1950 to 1952 when he retired. In 1953 Stewart was hired to make an inspection of the 1950-1953 section of the Alberta-British Columbia boundary survey. The photo was taken at that time. Gerry Smedley Andrews family collection

COMPLETION OF THE BOUNDARY SURVEY, 1950–1953

In his 1924 government report, Wheeler wrote about the uncompleted section of the 120th meridian north of 57°26': "The tract of country through which the meridian is unsurveyed lies in as yet virgin country, and there appears to be no immediate need for the establishment of the boundary." So Cautley's final station in 1923 marked the end of the survey of the 120th meridian for over twenty-five years.

But in 1949 circumstances changed, as described in the final Boundary Commission report.

During the following 25 years there was no pressing necessity for continuation of the boundary survey, but by 1949 two new developments made it a matter of considerable urgency. One was the undertaking by the Government of Alberta of a province-wide program of aerial mapping in connection with which it was found that the number of ground control points in the northwestern part of the province was inadequate to permit accurate plotting of topographical details from the aerial photographs; the other was the rapid extension of petroleum exploration into the northwesterly part of the province and beyond into the Northwest Territories and northeastern British Columbia. As far as Alberta was concerned, further demarcation of the boundary would provide the ground control points required for the aerial mapping work. It would also define on the ground the limit of jurisdiction between Alberta and British Columbia with respect to petroleum exploration.

Economic factors and jurisdictional considerations had also played an important role in the original Alberta-BC boundary survey. The procedure used to re-establish the boundary survey was also similar to the original, with the two provincial and federal governments all making formal requests, and each appointing a boundary commissioner. British Columbia's initial representative was its surveyor general, N.C. Stewart, who had worked on the original boundary survey during its final year. As was the case in the original survey, all three jurisdictions would share the costs equally.

The Boundary Commission report described changes made from the original surveying procedures.

> The northerly portion of the line is characterized by its straightness, the remoteness from the larger centre of settlement and the comparatively level terrain through which it runs. There is no possibility of dispute as to its position since the Imperial Act defines it as a line coinciding with the 120th meridian of west longitude. The task of establishing the meridian on the ground was relatively straightforward and could best be done by a single survey party. However, the Commission felt that the chief of the field party should be qualified as a land surveyor under the legislation of both of the provinces and preferably of Canada as well, and that the remoteness of this portion of the boundary called for a man with extensive experience on northern surveys.

The boundary commissioners held a preliminary meeting in January 1950 and appointed R. Thistlethwaite as its surveyor in charge of field operations. Thistlethwaite, from Saltspring Island, had surveying licences for all three jurisdictions. The Boundary Commission report noted: "From the outset it was apparent that the most important single factor in the progress of the work would be the matter of transport." In the field, the survey crew would find that Cautley's main obstacle still persisted. "It became the general rule during the present survey that the efforts of every member of the party, save perhaps the cooks, were directed part time to the clearing of the line in order to stabilize the general progress of the work which was, for the most part, dependent directly upon the rate at which the survey line could be cut through the woods." The boundary line had to have a skyline of 1.8 metres. "This, of course, necessitated a much wider clearing at the ground level; in fact it was often necessary to cut trees standing 15 or 20 feet [4.5 or 6 metres] from the centre line." Like Cautley, Thistlethwaite had a large crew of more than twenty composed of a second surveyor to operate the transit; a leveller and rodman to measure elevations; chainmen; a picketman; five axemen; a mound builder; a cook and helper; and four packers. Surveyor Knox McCusker, who had a ranch near Fort St. John, provided thirty horses and hired the packers.

In late May, the crew assembled at the Beatton River ford, about 50 kilometres from Fort St. John. The location was at the end of the road and the beginning of the old Donis trail, which headed north. The crew rafted the supplies and men across the river and swam the horses, then travelled for six days. They arrived at Monument 97-5, Cautley's final station, on June 9 and had no difficulty locating it. The Boundary Commission report stated:

> The days June 10 to June 13 were spent in retracing the sections of the boundary line between Monuments 97-3 and 97-5…. This period of retracement served as an ideal training and practice run for the alignment, chaining and levelling. At this time, contact was made with a seismic exploration party engaged in exploring the area east of the boundary for Imperial Oil Limited.

Beginning on June 13, the work of running new line continued without major interruption until September 19 when it was produced across the south crossing of Hay River and the terminal Monument 108-3 was set. Thistlethwaite and the crew surveyed slightly over 100 kilometres, and defined the boundary north of the 58th parallel.

The Boundary Commission report stated: "By good fortune, the season of 1950 was extremely dry. Had it been wet it is doubtful if the pack horses would have been able to carry out the necessary transportation for the survey party." The dry weather also had a negative result, for "fires were visible from the survey camp at all times. For a long time four distinct fires surrounded the party and were a source of considerable apprehension under certain wind conditions. It is very probable that some of the boundary surveyed this year was burned over after passage of the party."

In his book *The Chinchaga Firestorm*, Cordy Tymstra explained the survey crew's situation.

> Active forest fires were visible from the survey camps at all times during the entire 1950 boundary survey. Thistlethwaite reported being surrounded by four large fires for an extended period of time. This caused considerable anxiety for the crew as they ran ahead of and into fires. Some of the boundary line they cut traversed through recently burned areas, while other sections of the line burned after the survey crew moved north. Thistlethwaite and his crew threaded a burning needle and unknowingly survived one of Canada's worst firestorms.

The Boundary Commission report noted an unfortunate event during early August when

> a malady seized the crew and caused a high loss of time, particularly among the cutting force. The sickness is believed due to the use of stagnant water in what we have called Poison Creek [renamed Foulwater Creek by the Canadian Board on

Geographic Names].... The sickness disappeared when camp was moved away from this stream.

The surveying methods used were essentially the same as those used by Cautley.

As is customary, the boundary was projected across country as a series of successive extensions of the existing line. These are relatively short, ranging up to one-half mile or one mile [0.8 to 1.6 kilometres] in length. The forward point of each extension is normally determined by repetition of the well-known process of double-centering.

In accordance with the instructions, the direction of the line was controlled by periodic astronomic observation for azimuth.... The azimuth of the line was to be determined at intervals not exceeding six miles [9.6 kilometres] by a program of observation consisting of a minimum set of three observations on the Pole Star.

The Boundary Commission report noted that Thistlethwaite found no previous surveys to connect with his work. Most of the stations were marked with dirt mounds, but about every 10 kilometres, a concrete monument was established. This consisted of a standard surveying pipe post set in cement and gravel (if available).

The 1950 boundary survey employed some new techniques that had not been available to Cautley. The transits were more modern and accurate. Forestry radios enabled the crew to maintain outside contact. Trucks took the men and equipment partway to the site, and the crew (except for the packers) flew out at the end of the season.

Thistlethwaite decided to survey as long as possible and minimize the return time. Using radio communication, he made arrangements with Northern BC Air to fly his crew from Ekwan Lake, a three-day trip from their final survey station. In his government report, he wrote that on the journey a

matter of great concern at this time was the fact that a very large and briskly burning fire appeared to be directly in the line of march. It became apparent later that the party actually had travelled just under the nose, as it were, of the fire and that the trail was burned out a few days after the former's passage. Smoke conditions were severe throughout the trip.

The crew arrived at Ekwan Lake on September 22. Because of smoke and other reasons, Northern BC Air couldn't get a plane to the site. Thistlethwaite then arranged for a plane to come from Prince George and fly the crew to Charlie Lake near Fort St. John. Thistlethwaite wrote: "After one attempt to reach the party which aborted due to smoke conditions, the first landing was made on Ekwan Lake on September 30. This and the subsequent landing (October 1st) were made possible only by the use of radio communication between the aircraft and the field party, the latter 'talking him in' to the lake." He also noted: "By the time the aircraft reached the party, food had dwindled to nothing more than a little rice and macaroni, with occasional flavouring of spruce hen, rabbit, or jackfish."

There was a major change to the boundary survey in 1951, as described in the Boundary Commission report.

The nature of the terrain north of the point where Mr. Thistlethwaite's 1950 summer operations were terminated indicated that further progress with the survey during a summer season would be difficult and uneconomical, if not impossible. The territory which lay ahead consisted of a vast stretch of swampy country through which the line would have to be run for 20 to 25 miles [32 to 41 kilometres] before higher and firmer ground could be reached. It was advisable therefore to consider the possibility of carrying the surveying forward during the winter when frost conditions would render the ground passable.

In late November, the boundary commissioners asked Thistlethwaite to investigate the possibility of a winter survey beginning in January 1951, and he reported that he thought that it would be feasible to do so beginning in mid-January.

Although the method of surveying the boundary did not change, the structure did. Transportation would be mechanized and bulldozers would clear the survey line. Accommodation would be in portable housing that would be moved along the line as the survey progressed. Thistlethwaite's 1951 crew included no packers or axemen, but there were three men to work on the mounds and monuments, and five men to take care of the mechanical equipment.

The survey party's route to the boundary was via the Mackenzie Highway in Alberta to Upper Hay River, and from there on a well-used trail to Hay Lake Post. The trip was supposed to begin on January 22, but there were problems with some of the equipment, so part of the crew and equipment proceeded ahead. Most of the party departed from Upper Hay River on February 2 and reached Hay Lake, a distance of 100 kilometres, five days later. They stayed there until February 11, waiting for all of the equipment to arrive and making repairs. The crew reached the final monument of the 1950 survey on February 25. The Boundary Commission report stated: "The move from Upper Hay River, on the Mackenzie Highway, a distance of 145 trail miles [230 kilometres], had taken 24 days."

By now, Thistlethwaite had less than a month of surveying time available, for April 1 was considered the end of the winter season, and his crew and equipment needed to be back to Upper Hay River by then. The crew had a twenty-nine-day season and surveyed the boundary line 40 kilometres north, about half of what was originally expected. Part of this was caused by the long time it took to reach their location, and some of it was due to the different surveying procedures and learning to employ them efficiently.

The tractors moved camp and supplies daily. The Boundary Commission reported: "A notable feature of the work was the interference with survey operations by the presence of the tractors and sleigh trains in the cut line over which the survey was being made." When the bulldozers cleared the line they had additional work. "Inasmuch as the mobile camp had to be transported along the cut line… all stumps and trees had to be removed at the ground level.… Some blading had to be done also to smooth the way for them.… Creation of this trail provided good working conditions underfoot for the chaining and levelling parties." In the dense forest, the bulldozers could clear only about 1 kilometre a day. In the muskeg, the rate was much faster, but the risk of breaking through the ice and getting stuck was much greater.

The mechanical equipment was a vital part of the operation. It was important to properly use and maintain it to avoid breakdowns, and be efficient so that the equipment had minimal standby time.

The Boundary Commission report commented on the accommodation.

> The use of mobile housing results in a notable economy in the productive time of survey personnel. Owing to the fact that the living quarters were at all times within a mile or two [1.6 to 3.2 kilometres] of the actual surveying operations, unproductive walking time was minimized. Also, no time is spent in setting up and breaking camp or loading and unloading equipment as would be the case with a tent camp. Therefore, while the cabooses provided anything but comfortable living conditions owing to extreme restriction of space and poor regulation of temperature and ventilation, it seems that their use is advantageous.

Thistlethwaite had difficulty finding a transit that functioned well in cold weather, and he believed that the one he used had a defect. Thistlethwaite surveyed a large portion of the line at least twice, and his final results were satisfactory. He also had trouble with astronomic observations. In the Boundary Commission report, Thistlewaite stated: "In general, great difficulty was experienced in securing consistent agreement between the alignment and the astronomic controls. This led to excessive re-alignment and undue repetition of observations which in turn limited the overall progress of the survey to an appreciable extent. Astronomic work under low temperature conditions was found to be particularly trying, and there is no doubt that this circumstance contributed to the general difficulty." At times snow affected visibility, making surveying difficult.

The levelling and chaining crew and mound builders had to develop techniques that worked in cold weather. All of them eventually reached satisfactory results.

Although the 1951 survey did not achieve its objectives, the Boundary Commission gained much valuable information about surveying during the winter in a northern climate. The commission also realized that a winter survey would be necessary all the way to the 60th parallel.

Thistlethwaite was unavailable for the 1952 survey, so the Boundary Commission hired W.N. Papove, who also had surveying licences for all three jurisdictions and was on a survey project along the Alaska Highway during the summer and fall of 1951. Papove began preparations and consulted with several people. He decided to use a variety of transportation methods. Trucks would be used to haul equipment, supplies and horses to Hay Lake. They would transport only the supplies needed for the start of the survey, with the remainder flown in intermittently during the winter. The survey crew and some supplies would be flown to a site near the beginning of their work. Along the boundary, horses and sleighs would be the main source of transportation, supplemented by a dog team and a "motor toboggan" (snowmobile).

Papove intended to begin surveying early in January, but "the transport of men and supplies to the boundary was attended by several delays." In addition, "trucking operations between Hay Lake and the boundary were impeded by deep snow and efforts to speed up transport by conveying horse feed to the boundary by aircraft also proved to be impracticable." There were additional mishaps. The Boundary Commission report stated: "The extremely cold weather and deep snow also hampered activities and it was not until well into January that the survey operations got fully under way."

Papove decided not to use bulldozers to clear the line, but the main difficulty with the pace of the survey continued.

The main determining factor in the rate of progress of the field operations was the rate at which the line could be cleared. From six to eight men were available for cutting operations, but the rate of progress in clearing the line averaged only about one half mile [800 metres] per day. This resulted from the fact that line cutting operations had frequently to be suspended while the axe crews were diverted to clearing work on the sleigh road which had to be made for the movement of the camping equipment and which, for the most part, could not be suitably located along the boundary line.

For accommodation, Papove used large Norseman wall tents. Camps were spaced about 16 kilometres apart. Transportation of men and supplies required considerable organization because Papove had a crew of twenty-three people, and the survey line was continually moving. An airplane brought in more supplies about every two weeks, and the material had to be hauled from landing strips to camp. Papove had radio communication with the Hudson's Bay Company post at Hay Lake and the radio division of the Department of Transport at Fort Nelson.

Like Thistlethwaite, Papove had difficulties finding a transit that operated efficiently in cold weather. The Boundary Commission report noted: "For observing in temperatures ranging to -40°F [-40°C] enough emphasis cannot be placed on a careful and full winterization of instruments for such conditions."

Papove developed procedures to achieve acceptable surveying.

Difficulties were experienced in keeping a true north alignment of the boundary within the specified tolerance of 10 seconds of arc. In the production process great care was taken in setting backsights and making solid set-ups. Observations for azimuth on Polaris were taken at no greater than five mile [8 kilometre] intervals.

Papove also ensured that the chaining was as accurate as possible.

The survey proceeded uneventfully until late March, with the boundary delineated 59 kilometres farther north and past 59°N latitude. The men and equipment were flown out at the end of the season, while the horses and dog team returned overland to Hay Lake and the Mackenzie Highway.

Papove was unavailable in 1953, so for the final survey the boundary commissioners turned to a person familiar with the project—George Palsen, Cautley's leveller on the 1923 and 1924 survey of the 120th meridian. Palsen knew that in about 25 kilometres the line would cross the Petitot, the last river, and beyond there, the terrain would become more open and easy to cross. He recommended using dog teams for transportation and starting in November. Palsen hired several men who had previously worked on his surveys, and he arranged for transportation of men and supplies by air from Upper Hay River to an Imperial Oil landing strip about 40 kilometres from the boundary.

On November 8, eight dog teams, twenty men, and supplies arrived at Upper Hay River. By November 17, everything was assembled at a camp by the Imperial Oil landing strip. Many of the supplies were distributed by air to four caches near the boundary line farther north. It took eleven days to reach Papove's last station because of mild weather and insufficient snow. The Boundary Commission report stated: "The aspects in which this survey differed markedly from previous surveys were those of transport, terrain and weather conditions, all of which combined to promote much speedier progress." There were only a couple of cold weeks in January, and the snow was never deep enough to require snowshoes. "The ground over which the line ran was level or only gently rolling, the muskeg areas crossed were relatively small, and extensive tracts of forest growth had been recently burned, all of which contributed to easier opening of the line."

The survey of the line progressed at the rate of almost 2 kilometres a day and it reached the 60th parallel on January 21, after a survey of 80 kilometres. The southern terminus of the Alberta-BC boundary at Monument 272 on the US border was, appropriately

for the Rocky Mountains, located in a scenic col. The northern terminus, at the boundary with the Northwest Territories, was also appropriately located—in a small patch of muskeg.

During the next two days, Palsen connected the 120th meridian with the Alberta-Northwest Territories survey, which had placed their final station a short distance east of the line. Palsen then began surveying the British Columbia-Northwest Territories boundary along the 60th parallel and continued this work until March 9.

In the summer of 1953, N.C. Stewart, who had retired as BC's surveyor general, spent two weeks making an inspection of the 1950–53 portion of the boundary survey. He flew over the boundary and retraced a few of the stations done by each surveyor. Stewart declared his satisfaction with the surveying. Stewart also supervised the submission of the final plans for the survey. Another member of the 1918–24 boundary survey, Spike Thomson, prepared the material.

CONCLUSION

The results of the boundary survey showed that the project was successful. The most obvious and important physical delineation of the Alberta-BC boundary survey from 1918 to 1924 was the 120th meridian. The discrepancy of less than 6 metres over almost 200 kilometres between Cautley's survey from Pouce Coupe to Sheep Creek and McDiarmid's determination of the 120th meridian at that location showed that the boundary survey's line was accurate. The discrepancy in latitude was more sizable at 58 metres but still acceptable. The Boundary Commission report stated:

> The latitudes established by Mr. McDiarmid at Pouce Coupe and at Sheep Creek piers are astronomical, and are therefore subject to any deviation of the normal which may exist at either or both of these stations. As purely astronomic determinations frequently differ from those derived by geodetic methods by several seconds of arc, the above closings [less than two seconds] must be regarded as entirely satisfactory.

In his government report, Wheeler wrote about the connection of his survey and Cautley's, and its accuracy.

> From Torrens Mountain the meridian was carried south to its intersection with the watershed at Sheep Creek valley. Here a closing was made between the respective surveys of the two divisions. It is of interest to note that they closed satisfactorily, taking into account the difference of the methods and instruments employed, also the long stretch of country they have traversed, with starting points so widely apart and of such independent origins. In latitude the closing appears to be 143 feet [44 metres]; in longitude the closing appears to agree in the plot of the triangles. The distance between the two starting points in a straight line is approximately 214 miles [340 kilometres].... The altitudes carried respectively from determinations by precise levelling at Yellowhead Pass and at Pouce Coupe close to 11 feet [3.3 metres], a small matter in the consideration of mountain elevations and very nearly as close as the method of photo-topography can locate contours.

After the 1924 field season, Wheeler and Cautley spent several months completing all of the details related to the boundary survey and the three governments voted to ratify the work of the Boundary Commission. A permanent Alberta-BC boundary commission was established that would deal with maintenance of the boundary and any issues that might arise. This commission still exists, and the resurvey and remonumentation of the century-old stations from the original survey has been an ongoing issue, with work occurring in Elk and Yellowhead passes within the past five years. The 2017 fire that burned Akamina Pass will necessitate surveying there.

The boundary survey was the final project in A.O. Wheeler's surveying career. In the fall of 1924, he married Emmeline Savatard, who had been a friend of A.O. and Clara's for many years. In the next few years, Wheeler retired as president of the Alpine Club and gave up his involvement with the walking tours.

Part of a panorama of the Castleguard Glacier area taken in 1919. Repeat photography would show the extent of change in this vital watershed. v771_pd_21_72_001 and v771_pd_21_73_001, Whyte Museum of the Canadian Rockies

Wheeler spent his retirement travelling and doing a variety of outdoor activities.

In the fall of 1924, R.W. Cautley experienced personal tragedy when his wife died after an operation. But his surveying career continued. Cautley was hired by the Dominion Department of the Interior to a permanent position and moved to Ottawa. The following summer, he returned to western Canada as surveyor for the Alberta-Northwest Territories boundary survey. In 1927 and 1928, Cautley was the Dominion

representative for an investigation of suitable permanent boundaries for Banff and Jasper National Parks. (The Alberta representative was Lionel Charlesworth, the province's director of surveys during the first three years of the Alberta-BC boundary survey.) During the 1930s, Cautley examined several potential national park sites in the Maritime provinces.

After the boundary survey, A.J. Campbell began employment with the BC government's phototopographical survey. He joined his former surveying associate, R.D. McCaw, who had been with this branch of the BC government's surveying program since its inception in 1914. Campbell did phototopographical surveys throughout the province for about twenty years. In 1929, his surveying brought him back to the area that he had surveyed north of Mount Robson in 1923 and 1924. Campbell resurveyed some of the stations he and Lambart had established, along with Cautley's station on Intersection Mountain. Campbell surveyed the BC-Yukon boundary from 1946 to 1950.

In 1925, H.F. Lambart was part of the team that made the first successful climb of Mount Logan. Lambart had a successful career with the Geodetic Survey. In 1927 and 1928, he surveyed the north boundary of Jasper National Park for the Dominion government. His packer and guide was Felix Plante.

All three surveyors during the 1950 to 1953 completion of the boundary survey had successful careers. In 1953, Robert Thistlethwaite became surveyor general of Canada, a position he held for seventeen years. During that time, he was chairman for seven provincial and territorial boundary commissions. W.N. Papove surveyed sections of the BC–Yukon–Northwest Territories boundary (60th parallel) in 1953 and 1955.

The Alberta-BC boundary survey provided the first detailed maps of the terrain along the Great Divide of the Rocky Mountains. The physical evidence of the survey can still be seen in the monuments and along the cut lines through several passes along the boundary and on the 120th meridian.

The most important legacy of the Alberta-BC boundary survey is the approximately 5,000 photographs developed from the high-quality glass plate negatives that were taken as part of the phototopographic surveying of the Alberta-BC boundary. They provide full panoramas of the Rocky Mountain landscape as well as details of physical features and vegetation. These photographs provide an unparalleled opportunity to view the landscape of the Rocky Mountains as it was a century ago. Since they were taken at defined survey stations, it is possible to take pictures at the same location and examine changes in the Rocky Mountain landscape that have occurred since the boundary survey.

A large portion of the southern part of the boundary survey has been rephotographed. This rephotography, headed by Dr. Eric Higgs and the Mountain Legacy Project of the University of Victoria, has provided invaluable information to scientists in several disciplines. It will also provide reference points for rephotography projects in the future. However, much of the remote area north of Kicking Horse Pass, which includes numerous glaciers and icefields, has yet to be covered. It has been a century since the original photographs were taken along this section of the Alberta-BC boundary. The centennial years of the project are an opportune time to retrace some of the phototopography taken by the British Columbia crew between 1918 and 1924. It would provide valuable information for viewing landscape change in the Rocky Mountains, as well as establishing a legacy for future research. What this landscape will look like in the early twenty-second century is unknown, but a hundred year record, along with the original photographs, may provide indications for the ecology of these mountains and their priceless glaciers and watersheds. Scientists are expressing concern about downstream effects caused by the recession of the glaciers in the North Saskatchewan watershed. Rephotographing the glaciers along the Great Divide would provide information to assist them.

It is hoped that this book will be a catalyst for scientists to continue to expand the rephotography and to examine landscape change in the heart of this magnificent and important geographical region of Canada.

GEOGRAPHICAL NAMES

Since A.O. Wheeler did the topography for the Alberta-BC boundary survey, he had the opportunity to name the geographical features along the Great Divide. Wheeler subscribed to the "empty land" theory regarding the geographical features in the area he was surveying. Since very few of them had official names, he believed that his position as BC commissioner and his survey work along the Continental Divide gave him the right to name the features he mapped. Although the names had to be officially approved by the Geographic Board of Canada, it generally accepted Wheeler's recommendations.

Wheeler seldom consulted anyone about naming the geographical features, and he ignored the principle of naming peaks to reflect the natural or human history of the area. Unlike nineteenth-century scientists and explorers, Wheeler and his survey party rarely met Indigenous people who could provide names for the features of the Rocky Mountains, but he made no effort to consult them when making his maps.

The boundary survey occurred throughout World War I, and in a patriotic fervour, many of the geographical names Wheeler gave were related to the conflict. The first large set of World War I names approved by the Geographic Board of Canada began in December 1916. Most of these features named by Wheeler are located between Crowsnest Pass and Whiteman Pass, with the largest concentration in the vicinity of the Kananaskis Lakes–Elk Pass area.

Since the boundary survey spent almost all of its time in Howse and Yellowhead passes in 1917, Wheeler had no opportunity to submit new geographical names. But in 1918, north of Howse Pass, Wheeler surveyed several unnamed geographical features, and he once again gave many of them a name that had a World War I connection. This continued to a lesser extent in 1919, so between Howse and Fortress passes, there is a cluster of geographical features that have names related to World War I.

Wheeler was criticized for some of the names he selected in 1916 because they had no direct connection with Canada's involvement in the war. The names that he used in 1918 and 1919 are almost entirely battles in which Canada fought, or places where Canadian soldiers were stationed. Some of the mountains named for battles are Arras, Cambrai, Valenciennes, Fresnoy, Quéant and Zillebeke. Mons has an icefield, creek and peak named for this battle. La Clytte Mountain is named for a Canadian division reserve point. Almost all of the features Wheeler named are on the Continental Divide or in British Columbia. Willerval and Monchy mountains are entirely in Alberta. Mount Helmer is named for Alexis Helmer, a Canadian soldier who died during the war. John McCrae conducted his funeral, and this was the inspiration for his famous poem "In Flanders Fields." At the northern end of Wheeler's World War I geographical names there is a mountain, glacier, icefield and creek named for Georges Clemenceau, the prime minister of France during the latter part of the war.

Wheeler named Mount Alan Campbell along with Campbell Glacier and Campbell Icefield for his assistant. These features are in the area north of Howse Pass surveyed in 1918. Mount Jessie is named for Campbell's sister. It is located in Alberta, in the area north of Mount Robson surveyed in 1923.

The Alberta-BC boundary surveyors originally thought that the intersection of the Great Divide and the 120th meridian would occur around Jarvis Pass. In 1923, Lambart and his Geodetic Survey crew visited this area. During the fall of that year, Rev. George Kilpatrick, a member of Lambart's crew, submitted a list of names for geographical features in the Jarvis Pass area. It took over thirty years to resolve a controversy regarding one of these names.

Since Wheeler did most of the geographical naming, Kilpatrick wrote to him on November 30, 1923.

> Mr. Lambart has asked me to send you notes on the names which we gave to certain peaks and lakes in the district we visited this summer. I confess at the outset that we allowed our personal interests to determine our choice in practically every case. That may well rule out the names we selected, in the judgment of the Board, still it would be a rather difficult thing to name all the peaks in such a way as to have an obvious public interest. At any rate we may advance the claim that the names submitted do not offend against euphony and are not flippant.

Kilpatrick submitted a list of seventeen names.

On January 2, 1924, the Geographic Board of Canada wrote to Lambart that they had the sketch maps made by Mary Jobe and S. Prescott Fay and their names for geographical features around Jarvis Pass. (S. Prescott Fay and a small party, guided by Fred Brewster, had travelled between Jasper and Hudson's Hope in 1914. Fay's sketch map included the Jarvis Pass area and he named several features.) The board noted that several names that Fay submitted had already been accepted. "I may state that Mr. Fay is lucky in having these names accepted as the Board generally looks with disfavour on names suggested by tourists."

A week later, the board wrote another letter to Lambart stating that they would wait on selecting names until the boundary survey was finished. The letter also stated: "It is hardly in line with the Board's endeavour to incorporate the history of a region in the place names bestowed to approve for use as mountain names those of a host of Ottawa people who have had nothing to do with the mountains, a course of action to which the representatives of the provinces of British Columbia and Alberta would be certain to object."

In 1925, discussions started regarding names for geographical features in the vicinity of Jarvis Pass. It soon became apparent that there was some confusion regarding the names. On September 10, R. Douglas, secretary of the Geographic Board, wrote to Wheeler.

> In 1875 Jarvis and Hanington crossed Jarvis Pass as narrated in the Canadian Pacific Railway report for 1877. In 1916 the Board approved the name Jarvis for the mountain guarding the eastern entrance to the pass, named by Jarvis himself Smoky Peak from a mistaken idea that he had reached the head of Smoky River whereas it was the tributary of the Porcupine. This is apparently St. Andrews peak of the boundary survey [one of the 17 submitted by Kilpatrick and named for his church in Ottawa]. It is desired to name a peak on the north side of the pass after Hanington. Should this be the one marked Bronson on Sheet 39 or another further west?

In Wheeler's reply, he attempted to clarify the situation by referring to one of the phototopographical pictures that he had taken in the area. He pointed out the peak that he thought should be named Mount Hanington. (In contrast to Mount Jarvis, the naming of this mountain produced no controversy.) Wheeler also noted that he planned to be in Ottawa in November. Douglas replied that he would be glad to see him and "have your views on troublesome questions of nomenclature."

In October, discussions started regarding new names for the mountains on which the Geodetic Survey established stations. The mountain on which Lambart had a station named for his son, Arthur, was changed to Going Mountain, named for a Grand Trunk Pacific engineer who had worked in the area. Then it was changed to Fetherstonhaugh, named for another railway engineer who had worked in the area. Arthur soon received a third name change. "As Arthur peak is more outstanding than Boyd the Board would prefer that the name Côté [a senator from Edmonton] be applied to it and the name Fetherstonhaugh be

A view of Kakwa Lake with Mount Sir Alexander in the left background and Mount Ida in the right background. The geographical naming of features in this area proved contentious and took many years to resolve. v771_pd_59_101, Whyte Museum of the Canadian Rockies

transferred to Boyd." A nearby pass was also named Fetherstonhaugh. The mountain with the station named for Felix Plante was renamed DeVeber, for a senator from Lethbridge.

Discussions regarding names for geographical features continued throughout the fall. In early December, the Geographic Board informed Lambart that two of the seventeen names submitted had been accepted and noted that "St. Andrews is Mt. Jarvis of Board decision Feb. 1916." Nevertheless, Mount Saint Andrew's replaced Mount Jarvis.

In early January 1926, Lambart wrote to the Geographic Board: "In reference to names given to certain peaks in the vicinity of the Sheep Creek Pass on the Continental watershed... I beg to say that I have little further interest in the district more than to see the sheets well finished and the Topographical Surveys and our work well wound up."

After World War I, many geographical features across the country were named for soldiers who had been killed in the war. In January 1926, the Geographic Board wrote Cautley: "The Board is seeking a three-peaked mountain to name after three Mays killed in the war. I wonder if you can suggest one." Cautley replied that he could not think of one: "It seems to me that any distinctive mountain that is visible from some route of travel is a stupendous monument to any man or family, however gallant."

A.J. Campbell did phototopographic surveying for the BC government in the Jarvis Pass area in 1929. In 1932, as a result of his survey, there was a realization that Fay's sketch map had incorrect latitude and longitude for Mount Jarvis. The Geographic Board wrote: "This wrong latitude and longitude is confusing, so much so that Mr. Campbell, reading the note and believing it was some other mountain lying to the northward, gave another name to this peak. We have, however, corrected this."

There was also confusion regarding Mount Saint Andrew's. Campbell thought that it was named for the patron saint of Scotland and did not realize that it was the name of Kilpatrick's church in Ottawa. He named two nearby mountains Mount Saint George and Mount Saint Patrick for the patron saints of England and Ireland. Several years later another peak in the area was named Mount Saint David for the patron saint of Wales.

Though there was general agreement to reinstate the name of Mount Jarvis instead of Mount Saint Andrew's, it was never done. In 1957, the secretary of the federal Geographic Board wrote to the chief geographer of the BC government, re-examining the naming of Mount Jarvis. After reviewing the correspondence since the name had been adopted in 1916, he said that "there is little doubt that Mount St. Andrew's and Mount Jarvis are one and the same." S. Prescott Fay was still alive and a letter was sent to him suggesting that the name Mount Jarvis "be applied to the 8300 foot [2530 metres] summit south of Mount Saint Andrew's, unless the name was originally applied to the 6900 foot [2103 metres] summit north of Mount Saint Andrew's as shown on the enclosed map."

Fay wrote:

It seems a great pity to have so many of these early, very appropriate names changed to names that often have no special historical significance. The way it was prior to this last change, Mount Hanington was on the north side and Mount Jarvis on the south side of the exact height of land of Jarvis Pass; nothing could have been more appropriate.

In a letter sent in December 1957 to its executive committee, the Geographic Board stated: "Mount St. Andrew's was applied during the boundary survey and has become established on maps and in the gazetteer." The summit to the north was accepted. Though it is lower, it overlooks Jarvis Pass and the Jarvis Lakes. Fay concurred with the recommendation.

There were very few geographical features along the 120th meridian, so Cautley submitted only a few names. The ones that he selected along the mountainous southern portion of the meridian line were readily accepted. However, four names Cautley chose in 1921 sparked another dispute with British Columbia's surveyor general, J.E. Umbach.

The controversy started in October when Cautley replied to a letter sent by Deville.

I beg to acknowledge the receipt of your letter of the 5th instant, stating that the local names of natural features referred to in my general report of 1920 are objected to by the Geographic Board and asking me to suggest others.

All features referred to are of such a secondary character that in themselves they are scarcely worth naming in a district where very much more important features are still unnamed, except that it is desirable to name them for purpose of description in connection with the boundary map sheets.

Cautley provided four suggestions.

On December 12, 1921, Umbach wrote to Deville.

It has come to my notice that in connection with the work of the Inter-Provincial Boundary Commission that names are being applied to features in British Columbia which do not, in all cases, agree with existing names, and that apparently these names are being approved by the Geographic Board without any reference to the BC representative. I think it would be advisable that all names used in connection with the maps prepared by the Commission should be referred either to this office or to the office of the BC representative before they are officially approved. It might possibly simplify the procedure if these names were submitted to us direct in order that we might give same our approval or comments before they are submitted to the Board.

Deville replied: "I am unable to find any instances of the kind referred to. Could you give the names which you have in mind so that the matter may be looked into."

Umbach sent another letter on December 27.

In this connection I beg to quote the following extract from a letter dated December 6th by Mr. R.W. Cautley. It was the receipt of this letter which occasioned the writing of my letter of the 12th instant.

"I have had correspondence with the Geographic Board on the names Kelly Lake, Crooked Lake, Slough Creek and Muskeg Creek to all of which the Board objected. Of course it will be understood that I only used these local names on the returns of my 1920 season's work for purposes of identification.

The Board has seen fit to approve the following substitutional names for the above."…

I understood from the Board Secretary's letter that as all the above features either extend into British Columbia, or are altogether in it, you were being consulted in regard to the above nomenclature.

Cautley noted that three names were accepted and one was still under discussion.

Deville sent a letter to the Geographic Board. The secretary replied with a memo on January 3, 1922, in which he stated: "The rulings I gave were preliminary ones and it was always my intention to refer the names to the Alberta and BC governments when the boundary maps [underlined in original] were submitted to the Board." The secretary thought that Cautley's letter could have been phrased more accurately and in more detail. In summary, he stated: "1. The names complained of have not been submitted to nor approved by the Board. 2. All that was intended to be conveyed was preliminary approval of the alternative names suggested by the surveyor to replace poor names."

The next day Deville wrote to Umbach: "Enclosed herewith is a report by the Secretary of the Board from which you will see that the names in question have not yet been submitted to the Board and therefore have not been approved."

On January 11, Umbach wrote to Deville: "I regret that my letter of the 20th ultimo, written under misapprehension, has caused you so much inconvenience." This ended one of many disputes in which Umbach was involved.

SURVEY CREWS, 1918–1924

R.W. Cautley recorded the names of the men who worked on the Alberta crew every year. The names of the BC crew come from a variety of sources. The 1923 and 1924 BC lists are incomplete.

1918, Alberta

Cautley, R.W., chief surveyor

Sing, Mah Kai, cook

Gaucher, George, head packer

Plouffe, Fred, second packer (to July 21)

Maturen, Walter, second packer (July 22 to September 14)

MacPherson, John, monument-builder

Charlesworth, Gerald, chainman

Bayne, Lawrence, axeman

Mills, James, axeman (August 1 onwards)

1918, British Columbia

Wheeler, A.O., chief surveyor

Campbell, A.J., assistant

Thompson, A.E., second assistant

Martin, Tom, cook

Thomas, Dell, head packer

Hobock, W,C., second packer

Plymensen, Charles, second packer (August 7 to September 28)

Thomas, Guy, packer (August 8 to 17)

Cameron, Lee Grant

1919, Alberta

Cautley, R.W., chief surveyor

Carthew, J.T., assistant

Finnelly, C.M., cook

Ball, Frank, head packer

Clutton, Harold, second packer

Matson, Earl, third packer

Platz, Marcus, monument-builder

Charlesworth, Gerald, chainman

Henderson, Robert, picketman

Abernethy, Jim, axeman

Burnell, Arthur, axeman

Matthews, James, axeman

Pye, H.V., axeman

Reid, Ringland, axeman

Smith, James B., axeman

Souter, George, axeman

1919, British Columbia

Wheeler, A.O., chief surveyor

Campbell, A.J., assistant

Thomson, A.S., second assistant

Martin, Tom, cook

Rink, Ralph, head packer

Hughes, Arthur, second packer

Kain, Conrad

Nevler, Walter

1920, Alberta

Cautley, R.W., chief surveyor

Robertson, D.M., assistant

Maheu, Alex, cook

Ball, Frank, head packer

Matson, Earl, second packer (until August 8)

Burnell, Arthur, third packer

Laird, Robert, fourth packer
Platz, Marcus, monument-builder
Charlesworth, Gerald, chainman
McLeod, John, picketman
Gladue, Johnny, temporary guide
Elliot, Bob, axeman
Faulkner, Joseph, axeman (until August 10)
Gladue, Pascal, axeman
Gorsline, Alex, axeman
Hagan, Ole, axeman
Hartney, Chris, axeman
Helps, Cecil, axeman (until August 10)
McGregor, Thomas, axeman (until August 10)
Morley, Jack, axeman
Napoleon, Johnny, axeman
O'Morrow, Ross, axeman
Reid, Charles, axeman (until August 10)
Revell, Lance, axeman
Robertson, Edwin, axeman (until July 16)
St. Arnaud, Daniel, axeman
Thomas, Edward, axeman (until July 7)

1920, British Columbia
Wheeler, A.O., chief surveyor
Campbell, A.J., assistant
Thomson, A.S., second assistant
Weston, Fred, cook
Hughes, Arthur, head packer
McNair, E., second packer
Archibald, F.S.
Hunt, V.G.
Nevler, Walter

1921, Alberta
Cautley, R.W., chief surveyor
Robertson, D.M., assistant
Finnelly, C.M., cook
Condor, J.F., head packer
Gordley, H.H., second packer
O'Morrow, Ross, third packer

Platz, Marcus, monument-builder
Charlesworth, Gerald, chainman
McLeod, John, picketman
Burt, A.C., axeman
Voudray, Charles, axeman

1921, British Columbia
Wheeler, A.O., chief surveyor
Campbell, A.J., assistant
Moffatt, W.J., second assistant
Hough, Joseph, cook
Hughes, Arthur, head packer
Lamb, Jimmy, assistant packer
McKay, William, assistant packer
Stelfox, Jack, packer and cook
Archibald, F.S.
Archibald, R.W.

1922, Alberta
Cautley, R.W., chief surveyor
Gorman, A.O., assistant
Charlesworth, Gerald, instrument man
Neile, Harry, cook
Gordley, H.H., head packer
Bartholemew, Harry, second packer
Platz, Marcus, monument-builder
Nicholson, Tom, chainman
McLeod, John, picketman
Belcourt, Frances, labourer
Chaplin, Bob, labourer
Chaplin, Len, labourer
Cuthbertson, Ernest, labourer
Gilbey, Fred, labourer
Hollslander, Malcolm, labourer
Jordan, Jack, labourer
Napoleon, Johnny, labourer
O'Morrow, Ross, labourer
Thompson, Tom, labourer
Wilson, Ed, labourer

On August 12, 1922, Thompson, Cuthbertson, Jordan, Belcourt and Hollslander were discharged because they were no longer needed.

1922, British Columbia

Wheeler, A.O., chief surveyor
Campbell, A.J., assistant
Lee, Miles, cook
Lamb, Jimmy, head packer
Simpson, Leo, second packer
Bowman, P.E.
Nelles, Alex
Strong, R.D.

1923, Alberta

Cautley, R.W., chief surveyor
Gorman, A.O., assistant
Palsen, G., leveller
Charlesworth, Gerald, instrument man
Hunter, Peter, cook
MacKinnon, R.H., cook
Bartholemew, Harry, head packer
Leonard, Carroll, second packer
Napoleon, Johnny, third packer
Nicholson, T., chainman
Cramer, H.F., rodman
MacLeod, John, picketman
Miller, Dan, trail man
Baker, G., axeman
Brander, F., axeman
Gilbey, Fred, axeman
Neville, P., axeman
Simonson, G., axeman
Sturgis, Gus, axeman

1923, British Columbia

Wheeler, A.O., chief surveyor
Campbell, A.J., assistant
Thomson, A.S., second assistant

Smith, Charles, cook
Hughes, Arthur, head packer
Nejedley, Loren, packer
Bowley, Dan, packer

The names of three other members of the crew are unknown.

1924, Alberta

Cautley, R.W., chief surveyor

Phillipps Pass Survey
Charlesworth, Gerald, assistant
Whitcomb, Z., packer
Whitcomb, Paul, axeman

Summer Crew
Gorman, A.O., assistant
Palsen, G., leveller
Nicholson, T., instrument man
Belcourt, Frances, head packer
Gladue, Pascal, second packer
Neile, Harry, cook
Chaplin, R.
Howard, Sid
Shiels, Ed
Walker, S.D.

1924, British Columbia

Wheeler, A.O., chief surveyor
Campbell, A.J., assistant
Stewart, N.C., second assistant
Finley, Frank, packer

The names of the rest of the crew members are unknown.

Sources Consulted

Archives

BC Geographical Names Office: Geographical Names record cards and files

Geo BC, Surveyor General's boundary survey correspondence files:

    O-0008317, 1919

    O-0019671, 1920

    O-0027142, 1920–21

    O-0039110, 1922

    O-0045137, 1923

    O-0052276, 1923–24

Jasper Yellowhead Museum & Archives

    Jas1407, Arthur Hughes fonds

Library and Archives Canada

    RG21, registry records related to geographical names: Vol. 170, Vol. 173

    RG88, registry records related to the Alberta-BC boundary survey from the office of the Dominion surveyor general, E.G. Deville

    Vol. 153, File 17111, Cautley, 1920

    Vol. 153, File 17111, Cautley, 1920

    Vol. 154, File 17155, Wheeler, 1920–21

    Vol. 155, File 18020, Wheeler, 1922

    Vol. 155, File 18251, Wheeler, 1923

    Vol. 246, File 18500, Surveys, 1924

    Vol. 246, File 18501, Wheeler, 1924

    Vol. 364, File 16707, registry records, 1918–20

    Vol. 369, File 17137, Narraway, 1920–21

    Vol. 370, File 17524, Cautley, 1921–24

    Vol. 373, File 18001, registry records, 1921–23

    Vol. 373, File 18007, Narraway, 1922–23

    Vol. 374, File 17111, registry records, 1920

    Vol. 374, File 16898, Cautley, 1919–24

    Vol. 375, File 17372, reports, 1917–28

    Vol. 377, File 18002, registry records, 1921–23

    Vol. 381, File 18252, Cautley, 1922–24

    Vol. 381, File 18253, registry records, 1921–24

    Vol. 383, File 18502, registry records, 1923–25

    Vol. 397, File 20418, Lambart, 1922–35

    R1398-4-7-E, Howard John Frederick Lambart diaries 1922, 1923, 1924

    R1398-1-1-E, Howard John Frederick Lambart photo album

    1989-172 Geodetic Survey of Canada photo albums

Land Title and Survey Authority of British Columbia

    Boundary survey map atlases

Provincial Archives of Alberta

    A.S. Thomson fonds

Royal BC Museum and Archives

    AAAB5728, Felix Plante interview

    E/C/E31, Cautley, Robert. *Highlights of Memory* (memoir)

    MS733, Spike Thomson diaries, 1919–20

    198002-017, Vols. 7-8, R.W. Cautley, photo albums, 1918–20

Whyte Museum

    M546, A.O. Wheeler diaries, 1918–24

    V771, A.O. Wheeler, photo albums, 1918–24

BOOKS

Cotton, *Barry. Beating about the Bush: Random Memoirs of an Ex-Brit*. Victoria: Friesen Press, 2012.

Feddema-Leonard, Susan. *People & Peaks of Willmore Wilderness Park: 1800s to Mid-1900s*. Grand Cache, AB: Willmore Wilderness Foundation, 2007.

Fraser, Esther. *Wheeler*. Banff: Summerthought, c. 1978.

Helm, Charles, and Mike Murtha, eds.*The Forgotten Explorer: Samuel Prescott Fay's 1914 Expedition to the Northern Rockies*. Surrey, BC: Rocky Mountain Books, 2009.

Report of the Commission Appointed to Delimit the Boundary between the Provinces of Alberta and British Columbia. Part II: 1917 to 1921 from Kicking Horse Pass to Yellowhead Pass. Ottawa: Office of the Surveyor General, 1924.

Report of the Commission Appointed to Delimit the Boundary between the Provinces of Alberta and British Columbia. Parts III-A and III-B: 1918 to 1924 from Yellowhead Pass Northerly. Ottawa: Office of the Surveyor General, 1925.

Report of the Commission Appointed to Delimit the Boundary between the Provinces of Alberta and British Columbia. Part IV: 1950 to 1953 Latitude 57°26'40".25 Northerly. Ottawa: Office of the Surveyor General, 1955.

Sandford, R.W., and Jon Whelan, eds. *Every Other Day: The Journals of the Remarkable Rocky Mountain Climbs and Explorations of A.J. Ostheimer*. Canmore, AB: Alpine Club of Canada, 2002.

Tymstra, Cordy. *The Chinchaga Firestorm: When the Moon and Sun Turned Blue*. Edmonton: University of Alberta Press, 2015.

BRITISH COLUMBIA SESSIONAL PAPERS

1918: Campbell, A.J. "Vicinity of Swan Lake, Peace River District," N35–36.

1918: Wheeler, Arthur O. "Survey of the Boundary between the Provinces of Alberta and British Columbia," N84–88.

1919: Wheeler, Arthur O. "Survey of the Boundary between the Provinces of Alberta and British Columbia," G100–06.

1920: Wheeler, Arthur O. "Survey of the Boundary between the Provinces of Alberta and British Columbia," G130–35.

1921: Wheeler, Arthur O. "Survey of the Boundary between the Provinces of Alberta and British Columbia," H104–08.

1922: Wheeler, Arthur O. "Survey of the Boundary between the Provinces of Alberta and British Columbia," K113–20.

1923: Wheeler, Arthur O. "Survey of the Boundary between the Provinces of Alberta and British Columbia," F106–12.

1924: Wheeler, Arthur O. "Survey of the Boundary between the Provinces of Alberta and British Columbia," D123–28.

1950: Thistlethwaite, R. "British Columbia-Alberta Boundary Survey," H173–80.

1951: Thistlethwaite, R. "Winter Survey of Alberta-British Columbia Boundary," BB122–28.

Family Material

Alex Nelles family material: reminiscences

Jack Stelfox family material: written reminiscences

Geodetic Survey of Canada, Annual Reports

1922: Lambart, H.F. "Reconnaissance for Secondary Triangulation North of Yellowhead Pass, British Columbia-Alberta Boundary," 25–26.

1923: Lambart, H.F. "Alberta-British Columbia Boundary Triangulation," 24–27.

1924: Lambart, H.F. "Secondary Triangulation Alberta-British Columbia Boundary," 19–22.

Magazine and Newspaper Articles

Cautley, R.W. "Characteristics of Passes in the Canadian Rockies." *Canadian Alpine Journal* 12 (1922): 155–62.

Lambart, H.F. "The Canadian Rockies from Yellowhead Pass North to Jarvis Pass." *Canadian Alpine Journal* 18 (1929): 60–73.

"Some Verdict by Coroner's Jury." *Crag & Canyon* (Banff), July 17, 1920, 1.

Wheeler, A.O. "The Application of Photography to the Mapping of the Canadian Rocky Mountains." *Canadian Alpine Journal* 11 (1920): 76–96.

Wheeler, A.O., and H.F. Lambart. "Mountain Reconnaissance by Airplane." Canadian Alpine Journal 13 (1923): 146–57.

Wheeler, A.O. "Passes of the Great Divide." *Canadian Alpine Journal* 16 (1928): 150–76.

Websites

www.colwest.ca: "The 'Enigma' of the Lower Wales Glacier"

www.mountainlegacy.ca: Mountain Legacy Project

www.nrcan.gc.ca: Natural Resources Canada, R.W. Cautley diaries:

  FB18485: April–September 1921

  FB18486: September 1921–May 1922

  FB18959: May 1922–September 1922

  FB18995: April 1923–September 1923

  FB19444: April 1924–October 1924

INDEX